THE SOVIET UNION

THE SOVIET UNION

A NEW REGIONAL GEOGRAPHY?

Edited by Michael J. Bradshaw

Belhaven Press

London

HALSTED PRESS

WILEY

Halsted Press,

an Imprint of John Wiley & Sons, Inc.

New York Toronto

First published in Great Britain in 1991 by
Belhaven Press (a division of Pinter Publishers),
25 Floral Street, London WC2E 9DS

British Library Cataloguing in Publication Data

A CIP catalogue record for this book is available from the British Library

ISBN 1 85293 130 2 (hb)
1 85293 131 0 (pb)

Paperback edition only copublished in the Western Hemisphere by Halsted Press, an Imprint of John Wiley & Sons, Inc. New York Toronto [605 Third Avenue, New York, NY 10158–0012]

Library of Congress Cataloging-in-Publication Data

The Soviet Union : a new regional geography? / edited by Michael J. Bradshaw.
p. cm.
Includes bibliographical references and indexes.
ISBN 0-470-21815-0
1. Soviet Union – Politics and government – 1985- 2. Soviet Union -
-Geography. I. Bradshaw, Michael J.
DK288.S675 1991
914.7--dc20

91-29821
CIP

Typeset by The Castlefield Press Ltd., Wellingborough, Northants, England
Printed and bound in Great Britain by SRP Ltd., Exeter

Contents

List of figures

List of tables

List of maps

Notes on contributors

Brenton M. Barr is a Professor in the Department of Geography at the University of Calgary, Alberta, Canada. He is co-author, with Kathleen Braden, of the book *The Disappearing Russian Forest.*

Michael J. Bradshaw is a Lecturer in the School of Geography and an Associate Member of the Centre of Russian and East European Studies at the University of Birmingham, England.

Gary Hausladen is an Associate Professor in the Department of Geography at the University of Nevada-Reno, Nevada, United States.

Robert J. Kaiser is an Assistant Professor in the Department of Geography at the University of Missouri-Columbia, Missouri, United States.

Ronald D. Liebowitz is an Associate Professor in the Department of Geography at Middlebury College, Vermont, United States. He is Editor of the book *Gorbachev's New Thinking.*

Robert N. North is an Associate Professor in the Department of Geography at the University of British Columbia, Canada. He is author of the book *Transport in Western Siberia.*

Judith Pallot is a Lecturer in the School of Geography and an Official Student of Christ Church at the University of Oxford, England. She is co-author, together with Denis J.B. Shaw, of the books *Planning in the Soviet Union* and *Landscape and Settlement in Romanov Russia.*

Denis J.B. Shaw is a Lecturer in the School of Geography and an Associate Member of the Centre for Russian and East European Studies at the University of Birmingham, England. He is co-author, together with Judith Pallot, of the books *Planning in the Soviet Union* and *Landscape and Settlement in Romanov Russia.*

List of acronyms

ASSR	Autonomous Soviet Socialist Republic
BAM	Baykal-Amur Mainline
CDSP	Current Digest of the Soviet Press
CMEA	Council for Mutual Economic Assistance
CV	Coefficient of variation
FBIS	Foreign Broadcasting Information Service
FTO	Foreign Trade Organisation
FYP	Five-Year Plan
IMRs	Infant mortality rates
LDRs	Less developed republics
MDRs	More developed republics
OECD	Organisation for Economic Co-operation and Development
QR	Quintile ratio
TPC	Territorial-Production Complex
TsSU	Central Statistical Administration

Preface

My work as editor for this project has been punctuated by two visits to the Soviet Union, one in May 1990 and a second in May 1991. During the period between these visits a great deal happened in the Soviet Union: the 500-day Shatalin Plan was produced and later rejected, there were government reshuffles and resignations, there was threat of a civil war between Armenia and Azerbaydzhan, interior ministry troops repressed the independence movement in the Baltic, and more recently there has been the introduction of an 'anti-crisis plan'. After five years of unfulfilled promises, President Gorbachev is now very unpopular in the Soviet Union and even Western governments are questioning the sincerity of his reform plans. Soviet citizens have to cope with increasing taxes, the threat of hyper-inflation, diminishing incomes and rationing. However, in the early summer of 1991 there are again signs of renewed efforts to reform the Soviet economy: representatives of the republics and the central government are working towards a new Union Treaty, and new proposals are being drawn up to manage the transition to the market economy. President Gorbachev is currently seeking to persuade the major industrial powers to underwrite his programme for market reform. The election of Boris Yeltsin as the President of Russia is bound to challenge the legitimacy of President Gorbachev and the All-Union Government; at the same time, conservative forces in the central government are attempting to seize the initiative. Nevertheless, there is again hope that the Soviet Union can progress along the road towards a market economy.

The events of the past year convey both the excitement and frustrations of studying the Soviet Union today. New opportunities are available to travel to the Soviet Union and to meet with Soviet colleagues. At the same time, the pace and nature of the changes taking place in Soviet society often defy analysis: no sooner has a piece of research been completed than it is overtaken by events. However, the essays presented in this book aim to have lasting value because they examine the tension between the legacies of the past and the problems and prospects for reform. It is, admittedly, too early to write an authoritative 'geography of reform', in the sense of geography as the mere *outcome* of political and economic changes. However, any structural changes of this kind must progress within a framework of constraints and opportunities, and a crucial dimension of that framework is geography. The contemporary landscape of the Soviet Union remains, in large part, the landscape produced by seventy years of central planning and state socialism. It is, precisely, the right time to map out a 'new regional geography' of the Soviet Union, in which the legacy of stalinist space is highlighted as a critical context to the attainment of political and economic goals. The chapters review a variety of themes; in each, the authors identify the geographical consequences of the state socialist system, assessing the consequences for reform and, where it is visible, evaluating the initial geographical effects of the first five years of perestroyka (to the beginning of 1991).

The task of pulling together contributors in Britain, Canada and the United States has not been easy and I would like to thank all authors for the efficiency with which various scripts were sent flying across the Atlantic. The pace of change in the Soviet Union has meant that it is difficult to bring the process of manuscript preparation to any satisfactory conclusion. Any errors and oversights, therefore, remain the responsibility of the editor (and perhaps Mikhail Gorbachev!) I wish to thank Iain Stevenson and the staff at Belhaven Press for their support and assistance in this, my first attempt at editing a book. Over the past eighteen months a number of individuals and organizations have provided assistance. The Economic and Social Research Council provided two research grants which have supported my appointment as a Research Fellow in the School of Geography at the University of Birmingham. Financial support from the British Academy allowed me to attend the Association of American Geographers meetings in Toronto where two special sessions were devoted to this project. My two visits to the Soviet Union were supported by the British Council through an exchange between the Institute of Geography of the USSR Academy of Sciences in Moscow and the School of Geography at the University of Oxford. I am particularly grateful to Judy Pallot and Sergei Artobolevsky for arranging this academic exchange. I am also indebted to Andrey Treyvish, Leonid Smirnyagin and Alexei Novikov in Moscow and Viktor Lysenko in Novosibirsk for their help during my visits to the Soviet Union. Finally, I would like to thank Caroline for her support in the production of this book.

Michael J. Bradshaw
June, 1991
School of Geography
The University of Birmingham

1

Introduction: industrial restructuring and Soviet regional development

Michael J. Bradshaw

Introduction

This book represents the beginnings of a project to construct a new regional geography of the Soviet Union. The idea comes from a strong conviction that, in the process of societal transformation now underway in the Soviet Union, geography 'matters'. Just as space and geography play a fundamental role in the processes of economic restructuring in the developed capitalist world, so processes that are spatially constituted are an integral part of the attempted transition to a market economy in the Soviet Union. Geographers have a unique contribution to make in the analysis of reform in post-communist societies.

There are two senses in which the contributors to this book are constructing a 'new' regional geography. First, we consider the ways in which the contemporary regional geography of the Soviet Union constrains the progress of reform and the extent to which the spatial dimensions of reform are recasting the character and role of the different regions. Second, by incorporating the advances made by contemporary human geography, we aim to develop new ways of thinking about regional change in the Soviet Union. The geography of reform in the Soviet Union appears not as the outcome of Party decrees; but, rather, reform is the product of an interactive process between the central leadership and the various regions and republics.

Global restructuring and Soviet economic stagnation

It is now recognized that the late 1960s and early 1970s marked a profound change in the fortunes of the world economy. In the Western capitalist world terms such as 'post-industrial', 'post-Fordist' and 'postmodern' have been employed to describe the new patterns of economic and social organization that have emerged over the past two decades. The actions of the OPEC oil cartel and the subsequent dramatic increases in energy prices aggravated the economic problems facing industrial capitalism and triggered global recession. The socialist world was relatively isolated from the energy crisis; the energy resources of the Soviet Union enabled the Soviet bloc to avoid the economic stress caused by rapid increases in energy costs, while the Soviet Union itself won windfall profits from oil exports.

During the decade of the 1970s the major powers saw their industrial bases crumble as traditional manufacturing industries succumbed to rising costs, shrinking markets and increased international competition. Industrial geographers shifted their attention to the 'anatomy of job loss' as closure and rationalization became the dominant themes (Massey and Meagan, 1982). While traditional industries went into decline, the service sector became the source of growth and relative prosperity. The processes of decline and revival are spatially discrete: regions that have lost the traditional manufacturing jobs are not necessarily the same regions benefiting from the growth of service sector employment, and growth industries within the manufacturing sector are selecting 'new places' remote from the old industrial regions. The processes of industrial restructuring that have reshaped the developed capitalist world have also changed the nature of human geography. The work of geographers such as Massey (1984) and Storper and Walker (1989) has highlighted the crucial role of geography in the drive to accumulate.

The Soviet Union and the nations of Eastern Europe are now facing an economic crisis similar in scale to that experienced by the developed capitalist world during the 1970s. However, the single most important difference is that this economic crisis is paralleled by a wave of political revolution which has led to the downfall of communist parties in Eastern Europe and ended the central role of the Communist Party in Soviet society. The dramatic events of the recent past have shown that political change can be very rapid; unfortunately, it will take a considerably longer period of time to reshape the economies of the Eastern bloc.

In the immediate post-war period the Soviet Union succeeded in rebuilding its economy and expanding its manufacturing base. The extensive growth strategy developed under the dictatorship of Stalin was well-suited to the mass-production of standard products. The Soviet Union and East Europe mastered Fordist mass-production techniques on the grandest scale, the most obvious success indicator being the rapid growth of conventional armaments. Centralized control of the economy allowed the leadership to forge a military superpower at the cost of providing consumer goods and services for the population at large. To maintain growth rates, this extensive development strategy required more and more inputs of labour and resources. As long as resources were plentiful and cheap, growth could be sustained. But success indicators paid too much attention to simple plan fulfilment regardless of cost and quality. At the same time centralized control and pressure to produce more and more suffocated the innovation process. The system simply did not reward the introduction of new products or more efficient ways of producing existing products; any change in the production process meant risk and lost production.

The stagnating economy of the Soviet Union was controlled by a political élite incapable and unwilling, because of the political dangers, to introduce the radical reforms required. During the 1970s, when the capitalist world was experiencing economic crisis, the Soviet leadership under Brezhnev and Kosygin remained complacent. The energy crisis provided the Soviet Union with convertible currency to purchase grain and Western technology to bolster a faltering domestic economy. Increased East-West trade became a way of avoiding radical economic reform. Not surprisingly, the economic problems of the extensive growth strategy did not go away. By the end of the decade it became apparent that there was no imported technological fix; more significantly the cheap sources of labour and resources that had fuelled Soviet industry were no longer available. The energy base of the Soviet economy had shifted eastward into Siberia, and the centre of gravity of the labour-force growth had moved southward towards Central Asia. The Soviet Union was experiencing its own energy crisis as more and more investment had to be pumped into Siberia to fuel the domestic economy, supply Eastern Europe and generate foreign currency. The situation was made all the worse by the fact that the Western capitalist economies had weathered the energy crisis and by the mid-1980s were showing signs of recovery. New technologies and production processes were being introduced in the West that had no parallels in the Soviet economy. It no longer matters that the Soviet Union is the world's largest steel producer: the Soviet Union's once impressive industrial might is now a sign of relative backwardness.

When Mikhail Gorbachev became Chairman of the Communist Party of the Soviet Union in March 1985, he inherited a terminally ill economic system. In retrospect, the decade of the 1970s is now seen as an opportunity missed. Perestroyka was recognition that the economy requires radical economic reform, but glasnost has permitted the Soviet citizenship to call the Communist Party to account for previous mistakes. The processes of economic restructuring that have led to relative prosperity in the economies of the West have relied upon rapid technological change, entrepreneurship and decentralization. Soviet central planning had suffocated the innovation process and gave little or no power to the individual enterprise; the reforms introduced by the Soviet government have sought to accelerate technological change and encourage decentralized decision-making, but at the same time they are designed with the aim of maintaining some central control over the direction of the Soviet economy. However, it is now clear that the newly empowered Republican governments wish to follow their own trajectory towards a market economy. There exists no blueprint for such a transition from a command to a market economy and it should be no surprise that there has been limited progress to date.

Whether successful or not, the process of reform is changing the relationship between places in the Soviet Union and will, consequently, bring about dramatic changes in the patterns of regional development. Old and inefficient industries will have to be closed and new industries developed. At the same time as maintaining essential industrial sectors, such as energy and transportation, the economy will have to shift emphasis away from the military and heavy machine-building sectors towards consumer goods and services. At the republican and regional levels economic activity will have to be reoriented to local, rather than national, needs. Some regions will benefit from new investment and opportunities, while others will suffer from deindustrialization. No doubt the inhabitants of those regions that stand to suffer will criticize the transformation to a market economy; similarly those regions that are in the vanguard of reform, such as the Baltic republics, are using this opportunity to push their claims for independence. This process of regional differentiation takes on new meaning when laid against the complex cultural landscape of this multinational state. Perestroyka and glasnost are not only changing the geography of the Soviet Union, but are allowing it to be expressed in nationalism, regionalism and environmentalism. Geographical diversity is at the centre of the centrifugal forces that appear to be pulling the Soviet empire apart. To date, the central authorities have failed to recognize the importance of the geographical dimensions of reform, a factor that goes some way to explaining the leadership's inability to deal with the rising tide of separatism.

Regional geography and perestroyka

If there are similarities between the intent of Gorbachev's perestroyka and the processes of industrial restructuring experienced in the developed capitalist economies of the West, then there must be theoretical and conceptual implications for the way in which we piece together our analyses and interpretations of industrial change and regional development in the Soviet Union.

Industrial geographers in the West have sought to explain patterns of regional economic development as the consequence of interaction between large-scale global and national forces, and regional- and local-scale conditions. The foundation of 'restructuring theory' (Lovering, 1989) is Doreen Massey's concept of the 'spatial division of labour', which depicts the interdependent but specialist functions played by economic activities in different regions (Massey, 1984). Each cycle of accumulation produces spatially uneven effects, as the search for profitable locations for new industrial functions responds to the enabling and constraining qualities defined by the 'historically prior social uses of space' (Warde, 1985, p. 195). In the pursuit of profit, technological change and the processes of creative destruction forge a dynamic landscape of opportunities, resulting in the persistent restructuring of economic activity as capitalist enterprises (particularly multinational corporations) seek out new opportunities. The metaphor employed by Massey (1984) is geological, likening each major round of investment in an economy to a phase of sedimentation: the current surficial features of deposition are shaped by pre-existing layers which are themselves products of previous periods of deposition. Thus, the present landscape reveals its past and helps shape its future. The strength of the 'spatial division of labour' approach is that it links regional and local development to processes operating at broader scales. For geographers studying developed capitalist economies it is the logic of capital accumulation that promotes new rounds of investment. Governments and economic organizations at all scales seek to influence this process by attracting and directing investment or by erecting barriers against the entry of undesirable activities.

Increasingly, studies of industrial restructuring and regional development have sought to place local-scale change within the context of national policy and global processes. The most concrete expression of this new concern in Britain has been the project of so-called locality studies (Cooke, 1989). This has been associated with an apparent 'revival' of regional geography as a means of interpreting and explaining the local impact of restructuring. Calls for this 'new regionalism' stem from a theoretical imperative to bridge the gap between structure and agency. However, a considerable distance remains between the advocates of 'new regional geography' and the kind of contemporary regional specialisms that the more conventional 'regional geographer' now pursues (Bradshaw, 1990). This gap is unfortunate: both approaches have something offer. What good is theoretical sophistication without detailed regional knowledge to flesh out explanation and provide the essential ingredients of place? Similarly, detailed regional descriptions that fail to relate local change to wider processes lack explanatory power. What is required is a 'reconstructed regional geography' that combines empirical expertise with theoretical sophistication.

I want to suggest that, as we seek to examine the spatial consequences of reform in the Soviet Union, we can learn from the approaches taken by those who have been exploring the spatial dimensions of restructuring in the Western capitalist economies. I am *not* suggesting that the processes of industrial restructuring are the same in the Soviet Union – clearly they are not – but that concepts such as the spatial division of labour can be marshalled to the cause of analysing the transformation of State Socialism in the Soviet Union and Eastern Europe. Undoubtedly there will be those people who will have problems with the application of structuralist concepts to the Soviet Union, which in itself raises interesting questions about the nature of Western Capitalism and State Socialism. But, in times such as these, terms such as 'Capitalism' and 'Socialism' seem drained of explanatory power (Lane, 1989). Part of our problem results from the tendency of Western geographers to see industrial restructuring as synonymous with Capitalism. From the point of view of the Soviet specialist it is more profitable to view Western Capitalism as one type of industrial society and Soviet State Socialism as another (Davis and Scase, 1985). Both types of industrial society share technological processes: it is the legal ownership and social organization of production that differ. We may find it appropriate to use some of the concepts developed by Western industrial geographers because they address the impact of technological change upon the production process – and indeed the acceleration of technological change is at the very heart of perestroyka. Moreover, in both systems surplus value is extracted from the workers: in the Capitalist System it is the owners of the means of production who reap the benefits of accumulation, whereas in the State Socialist System the state collects and allocates the surplus on behalf of society. (In practice the Soviet state has used the surplus to promote military strength through heavy industrialization rather than maximizing the benefits to the populace.) The current crisis in the Soviet Union is in a very real sense a crisis of capital accumulation, and the State is having to manage an increasing fiscal deficit. We find uneven development in the Soviet Union because Soviet leaders and planners also recognized the contradictions between equality and economic growth, and in general have sided with economic growth in the guise of fulfilling national economic plans.

Restructuring and Soviet regional development

Recent developments in human geography have not only declared that geography 'matters', but have also stressed the importance of history in understanding the contemporary landscape. The historical legacy is equally crucial to understanding current regional problems in the Soviet Union. The Soviet Union, and before it Tsarist Russia, experienced technological cycles similar to those of the industrialized West. The differences between the Soviet economic system and the market economies of the West influenced the pace and pattern of restructuring. First and foremost, the competitive pressures of the market did not exist in the official Soviet economy: rather, economic development and technological change had to be directed from above. Second, non-economic factors such as ideological, strategic and political considerations shaped the structure of economic activity in the Soviet Union more directly than in market economies (Schiffer, 1989). Third, the Soviet economy has been relatively isolated from the competitiveness of the international market place; however, this does not mean that the Soviet Union has been completely removed from the pressures that promote technological change and, as we have already noted, the Soviet economy needs to accumulate capital to fund economic growth.

Prior to the industrial revolution in Europe, the Urals Industrial Complex, developed by Peter the Great in the early eighteenth century and based on local iron ore and forest resources to provide charcoal, was one of the world's major industrial metallurgical centres (Blackwell, 1982, p. 20). By 1763 the Urals produced two-thirds of Russia's iron in factories, and Russia exported iron ore to England. However, the industrial revolution in Britain soon relegated the Urals to serving a highly protected domestic market. In his examination of the industrialization of Russia, Blackwell (1982) identified two major industrialization drives, the first between 1856 and 1913, and the second between 1929 and 1953. Goldman (1987) has likened Gorbachev's program of perestroyka to a third industrial revolution.

The first industrialization drive in the second half of the nineteenth century, although somewhat delayed, replicated the industrial revolution that swept across Western Europe and North America. As elsewhere, the leading sectors were associated with the harnessing of steam power; the key locational requirements included the availability of iron ore and coal. Places such as northern England and the northeast of the United States were emerging as major centres of economic activity and, similarly, in Tsarist Russia the focus of new activity centred around the coal fields of the Donetsk Basin (Map 1.1) and the iron ore of the Krivoy Rog. At the heart of this emerging industrial geography was the utilization of new sources of energy, new technologies and new patterns of settlement (Blackwell, 1983, p. 388). Regions such as the Urals fell into relative decline, while areas such as the Ukraine gained in importance. On the eve of the First World War the Ukrainian region accounted for 74 per cent of Russia's total pig iron production of 4.2 million tons, while the charcoal furnaces of the Urals accounted for 22 per cent (Shabad, 1969, pp. 35–7). The spatial impact of this industrialization drive was felt mainly in the European and Trans-Caucasian regions of the Russian empire; the lands east of the Urals and those newly acquired in Central Asia were relatively unaffected.

Further consolidation of the economic geography of Tsarist Russia was cut short by the First World War, the Revolution and the Civil War. Lenin's New Economic Policy sought to re-establish industry, but it was not until Stalin's First Five-Year Plan in 1928 that a concerted effort was made to rejuvenate the industrial economy of the Soviet Union. The creation of a command economy and the guiding role of Marxist-Leninist ideology obviously shaped the nature and pattern of Stalin's industrialization drive. The Soviet Union not only had to restore the damage done by years of struggle, but it also had to try and catch up with the technological advances that had been made in the West. Stalin's policies served to consolidate the dominance of the European regions, but at the same time ideological and strategic concerns promoted the industrialization of peripheral regions in Siberia and Central Asia. By

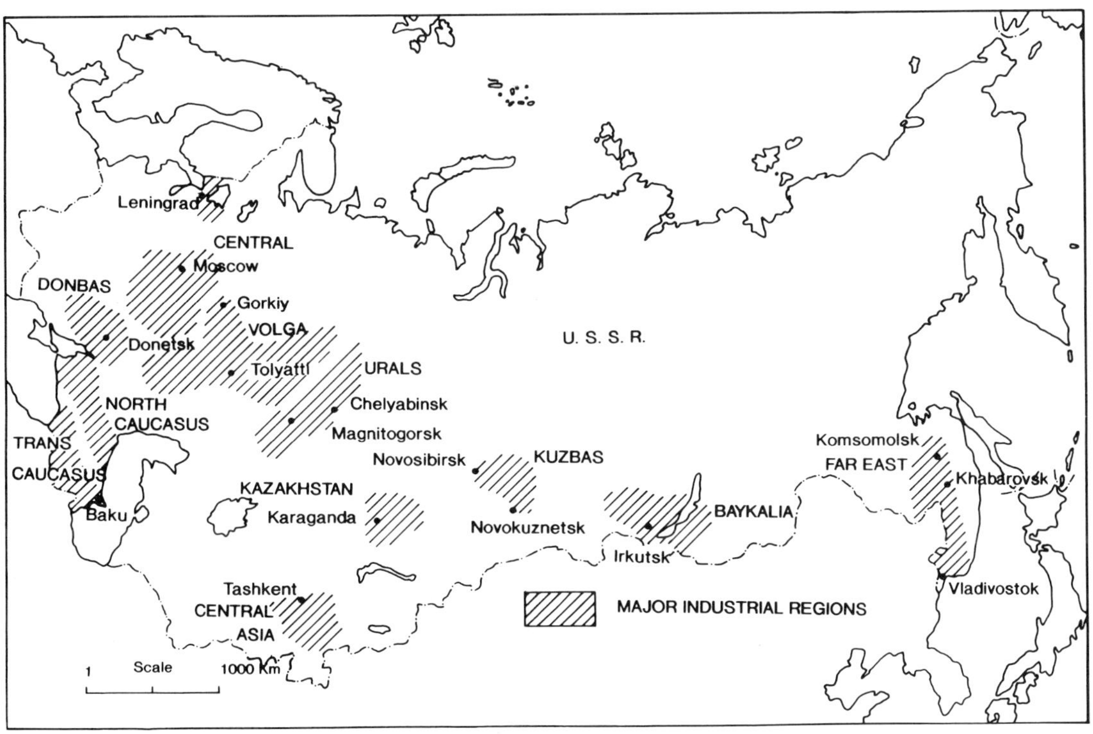

Map 1.1 *Major industrial regions in the Soviet Union*

the early 1930s new heavy industrial centres were emerging, such as Magnitogorsk in the Urals, Stalinsk (Novokuznetsk) in West Siberia, as well as Zaporozhye, Tula and Lipetsk in the European regions of the country. The Urals-Kuznetsk Combine re-established the Urals as a major metallurgical base. Coal production was increased in the Urals, Karaganda (Kazakhstan) and the Kuznetsk basin. Rapid industrial growth and urbanization came at the expense of rural deprivation; forced collectivization extracted capital from the countryside to fund the construction of steel works and factories. The infrastructure of the cities was swamped by an influx of workers, while a reign of terror in the countryside led to widespread starvation as many farmers destroyed their crops and livestock rather than hand them over to state authorities. There was an eastward drift in the location of industrial activity, but the European regions of the Soviet Union still contained the economic core. An indication of the dominance of the European regions is provided by the fact that the areas occupied by German forces up to November 1941 had been responsible for 63 per cent of Soviet coal production, 68 per cent of pig iron production and 58 per cent of steel production (Nove, 1982, p. 271). Ironically, the Second World War served to promote the economic development of the peripheral regions of the Soviet Union as many industrial enterprises were relocated in the Urals, West Siberia and Central Asia.

The Second World War imposed further setbacks upon the Soviet economy and the immediate post-war years were spent reconstructing the industrial base of the European regions. Heavy industry still dominated, but the conversion from solid to liquid fuels exerted influence upon the location of productive forces (Figure 1.1). The 1950s were marked by the rapid development of the Soviet petroleum industry. Prior to the Second World War the Soviet petroleum industry was concentrated in the Caucasus, where commercial production at Baku dated from the 1870s (Shabad, 1969, p. 12). In 1940 oil accounted for 18.7 per cent of Soviet fuel production, and by 1960 its share had risen to 30.5 per cent. As oil production increased and new areas came into production, so the geography of energy production changed. By the mid-1960s the Volga-Urals field contributed about 70 per cent of Soviet crude oil production (Figure 1.2). The emerging petrochemical industry was not initially drawn to the oil-producing regions; instead new refineries were located in consuming areas and supplied by pipelines. Thus despite the inadequacies of the resources base, the European USSR was able to maintain its dominance by transporting the necessary resources from other regions. The geographical spread of oil (and later gas) production prompted the development of pipeline systems

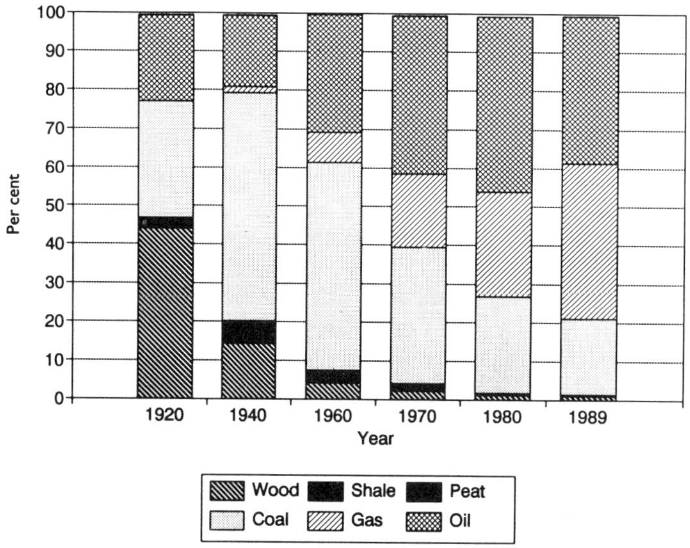

Figure 1.1 *Changing Soviet energy mix: per cent of energy resource production*

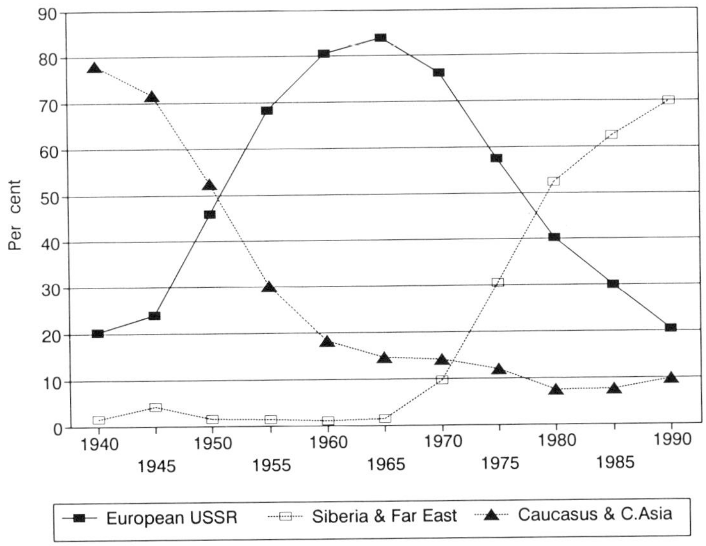

Figure 1.2 *Macro-regional distribution of Soviet oil production: 1940–90 plan*

which continues to this present day. At the same time the development of resource sectors in Siberia and Central Asia placed a great stress upon the transportation system. Within the European core region there was an eastward drift as new manufacturing industries, such as the automotive and petrochemicals industries, were developed in the Volga-Urals area. By the late-1960s a core-periphery pattern of development was well-established. The bulk of industrial activity, as well as the market, was concentrated within the European regions, while the Siberian and Central Asian peripheries provided natural resources. This pattern of regional specialization was not the consequence of an explicit regional development policy: rather it was the spatial outcome of the combination of sectoral policies. With the exception of Khrushchev's *Sovnarkoz* reforms, the Soviet economy has been planned by industrial sector rather than region. As the Stalinist extensive growth strategy demanded more and more labour and industrial raw materials, it forced the resource frontier into ever more remote regions, depriving the countryside of a much-needed workforce in the process.

Production from the Volga-Urals oilfields peaked in the mid-1970s and Soviet geologists looked to West Siberia for future oil supplies. In 1970 the West Siberian economic region supplied 8.9 per cent of Soviet oil production; in 1989 its share was 67.3 per cent (408.5 million tons). During the late-1970s the Soviet economy, in common with the industrialized West, experienced a shift from oil to natural gas as the favoured resource. Again the peripheral regions of Siberia and Central Asia have provided the bulk of natural gas production. By 1989 only 13.2 per cent of Soviet natural gas production originated in the European regions of the country. These dramatic shifts in oil and gas production have not been paralleled by shifts in the chemical industry. In 1970 the European regions of the USSR accounted for 86.8 per cent of the gross output of the chemical industry, with Siberia contributing 9.3 per cent; in 1985 the relative shares were 84.9 per cent in the European regions and 9.5 per cent in Siberia (Sagers and Shabad, 1990, pp. 42–3). Despite recommendations by Siberian economists (Granberg, 1986), only a modest amount of resource-intensive industry has been located in resource-producing regions. Today the geography of the Soviet Union reflects the spatial division of labour, and the consequent pattern of regional specialization, generated by Stalin's industrialization drive. Satisfying the resource demands of the European core region has been a costly exercise. During the 1960s one rouble of production in the extraction industry required two roubles of capital investment, but during the 1981–5 period the same rouble of production required seven roubles of investment (Aganbegyan, 1988, p. 71). During the

1990s the cost of resource production will continue to increase as the resource frontier is pushed further northwards in the Arctic and offshore into the Barents Sea and Sea of Okhotsk.

At the same time as Soviet planners were spending billions of roubles developing the empty taiga and constructing transcontinental pipeline systems, demographic trends were exacerbating labour shortages in the European core region (Map 1.2). Meanwhile there were increasingly large numbers of idle hands in Soviet Central Asia. Stalinist policies promoted a dependent relationship between the Central Asian regions and the European core (McAuley, 1985, p. 307). The economy of Soviet Central Asia has been developed to supply industrial agricultural crops, particularly cotton, for industries in the European regions; consequently the region has had to rely on outside sources of manufactured goods and food supplies (Dienes, 1987, pp. 122–33). The key issue in Central Asia remains how to utilize the rapidly-growing labour force. Kingkade (1987, p. 175) has estimated that between 1991 and 1995 the Central Asian nationalities will account for 80.4 per cent of the growth in the working age population of the Soviet Union. At the same time, the European regions of the Soviet Union are experiencing low population growth rates and suffer from a shortage of labour. The regional problems of the extensive development strategy are best summarized as a spatial mismatch of the factors of production: industry and market located in the European regions, additions to the labour force in Central Asia and natural resources in Siberia and the Soviet Far East. In turn, each macro-region has its own particular set of geographical problems.

By the late 1970s the extensive mode of development and its pattern of regional interdependence had served to put a brake upon the growth of the Soviet economy. The Stalinist economic system had failed to promote technological innovation within the leading sectors of the economy. Economic growth could no longer be obtained by consuming more labour and more resources, for the necessary labour was no longer available in the European regions and the cost of additional resources was becoming prohibitive. Today the Soviet economy suffers under an increasingly dysfunctional spatial division of labour, and only radical economic restructuring can breathe new life into it.

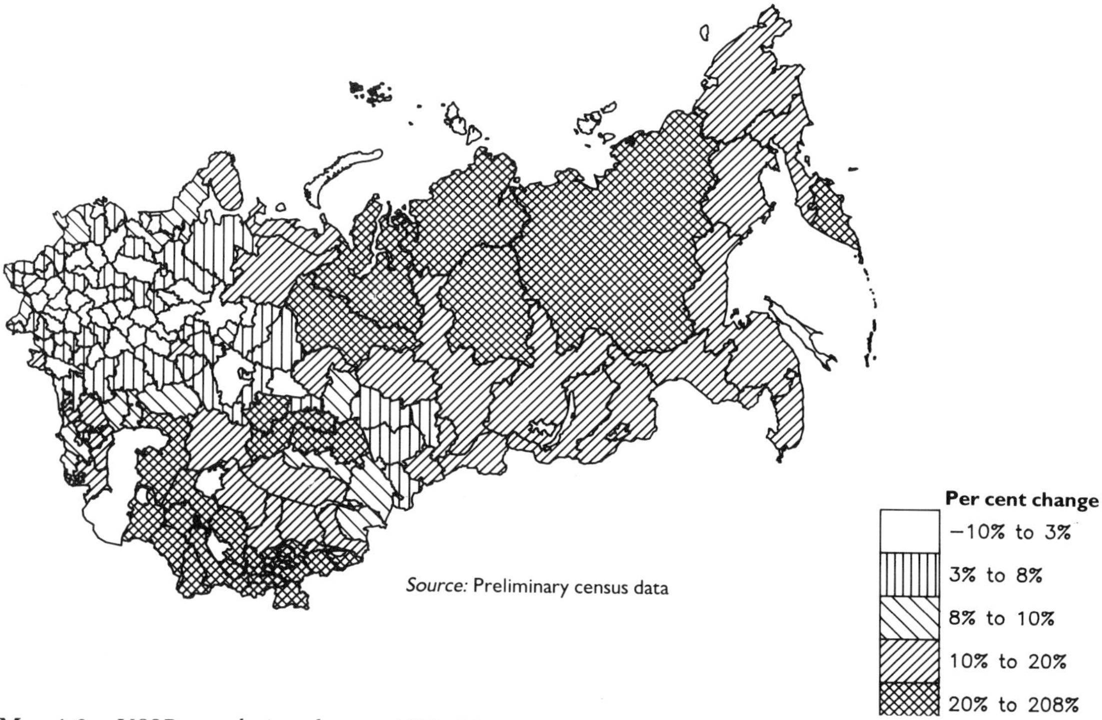

Map 1.2 *USSR population change, 1979–89*

Perestroyka represents an attempt to promote an intensive growth strategy, emphasizing efficiency, technological innovation and conservation. What we are currently witnessing is the dismantling of the Stalinist extensive growth strategy that enabled, at considerable human cost, Soviet industrialization. With the present accent on stagnation, however, it is worth noting that the Soviet period (and particularly the post-war period) has seen considerable geographical change (Lewin, 1988). Even during the Brezhnev-Kosygin period there were dramatic changes in the geography of economic activity and population distribution, many of which have contributed to the economic problems currently facing the Soviet Union. The investment strategy of the current policies aims to produce a technologically advanced, competitive market system capable of satisfying the need of the people and maintaining the Soviet Union's status as a superpower. The first five years of perestroyka suggest that this will be no easy transition and that the Soviet state may not survive intact.

Interpreting perestroyka

There is a school of thought that it is premature to speak of a geography of perestroyka, for we will have to wait at least ten years before there can be concrete changes in the economic landscape of the Soviet Union. The geographer's task, according to that view, will then be to describe the geography of this new reality. Such a perspective points to a key distinction between traditional regional geography and the intentions of new regional geography. For the traditional regional specialist, the principal goal has been to describe the outcome of development in terms of patterns of economic activity, settlement, population and so on. In the case of Soviet geography, problems of language, accessibility and data availability have made even 'simple' description a difficult task and I do not doubt the value of digging out and conveying information. However, traditional regional specialists often failed to examine the spatial constitution of the processes that underlie those patterns being described (a notable exception is Pallot and Shaw, 1981). All too often geographers (in East and West) have been happy to describe outcomes and have sought little explanation beyond five-year-plan statements (an approach that is no longer available to us). As a consequence, within Soviet studies, geography has been assigned a passive role as a collector, collator and presenter of 'facts'. Current events in the Soviet Union demand that we adopt a more active analysis of the spatial dimensions of reform. Indeed, without taking this dimension into account, the interpretations constructed by other social scientists lack a perspective that is crucial to explaining the progress of perestroyka.

This scholarly error mirrors a similar shortsightedness in the organization of the Soviet economy itself. Despite the numerous reforms aimed at decentralizing decision-making the Soviet Union is still planned, and reformed, by economic sector. To date, economic reform has addressed specific sectoral problems such as poor agricultural performance or inadequate provision of consumer goods, rather than regional problems such as under-employment in Uzbekistan. At present, there exists no regional strategy that is a component part of perestroyka, although there is now talk of a regional development fund. However, this does *not* mean that there is no geography of perestroyka: on the one hand, decisions made in its name have a spatially-differentiated impact upon the landscape; on the other hand, the regions and peoples that occupy the landscape respond differently to reform. Reform is an interactive process between the politicians and strategists in Moscow and the governments and people of the constituent republics. Quite clearly one of the problems at present is that the populace has little faith in the strategists in Moscow! A failure to account for the geographically differential impacts of reform may be vital to understanding opposition to the 'new economic reality'. The differential impact of sectoral restructuring has major repercussions for regional development: for example, an investment strategy that favours the modernization of the machine-building industry also favours the European regions of the Soviet Union simply because that is where 90 per cent of the machine-building industry is located. At the same time, by default, such a strategy does little to solve economic problems in Central Asia or Siberia (Bandman and Linge, 1991). Regional variations in the creation of joint ventures, for example, reveal the Baltic republics to be clear winners, while traditional industrial regions such as the Donbass and Kuzbass are relative losers. As the centre loses legitimacy, regional interests are challenging the Soviet government; for example, oil workers in Tyumen have used strike threats to press for better living conditions. It is important to identify where these economic winners and losers are since they have implications for political cohesion.

In sum, it is necessary to trace the consequences of and reaction to sectoral reform across the landscape of the Soviet Union. In the past, patterns of Soviet regional development were little more than a collage of various sectoral decisions made in Moscow, since the region was often a passive recipient of investment from the centre. Perestroyka and glasnost have provided an opportunity for local response to national strategy. The ability of the various republics and regions within the Soviet Union to seize the initiative (even to the extent of declaring independence) and make the most of the opportunities offered by reform has become an important factor in explaining contemporary patterns of regional development. However, there are disturbing signs that the centre is unwilling to relinquish control over economic and political matters. During 1990 a battle of wills, and laws, emerged between the central All-Union Government and the various Republican Governments. It remains to be seen who the eventual victors will be, but resolution of inter-governmental conflicts must be a prerequisite for the introduction of further reforms.

Research agenda

There is an element of truth in the view that it is a little early to *describe* a fully-fledged geography of perestroyka. When Mikhail Gorbachev came to power in 1985 he did not inherit a blank map, and the present geography of the Soviet Union reflects past strategies, not perestroyka. The task in this book is to identify the problems inherited from and created by the Stalinist extensive development strategy, to evaluate the policies proposed to solve past problems, and to consider the spatial consequences of central sectoral decisions in combination with local responses to national strategy.

This text makes no claim to be comprehensive, an impossible task given the current pace of change. It reflects the interests of a group of geographers who are concerned to develop a new regional geography of the Soviet Union. The book opens with an examination of the issue of spatial inequality. In 1972 Leonid Brezhnev declared the problem of regional equality to be solved; since then both Soviet and Western studies have refuted this claim, and there remains considerable variation in living standards between the various republics and regions of the Soviet Union. In Chapter Two Ronald Liebowitz documents the difference between rhetoric and reality and considers how perestroyka might change degrees of inter-regional inequality. In Chapter Three, Robert Kaiser explores the ethnic dimensions of spatial inequality. The rise of nationalism has been one of the most disturbing aspects of change in the Soviet Union, since glasnost has provided the opportunity for various ethnic groups to voice their discontent. In the Baltic this has led to calls for independence which have evoked a violent response from the Soviet government, while in the Transcaucasus nationalist tensions threaten the very future of the Soviet State. How are these nationalist demands to be assessed? Is their solution to be found in a new Union Treaty that reinterprets Soviet federalism, and can the Soviet Union benefit from the experiences of other multi-ethnic states?

Issues of spatial inequality tend to focus upon various regions within a state, or between states. In the Soviet Union some of the most stark variations in living standards are to be found between urban and rural settlements. Stalin's industrialization drive, and the central planning system that orchestrated it, prompted rapid urban growth and shaped the form of the Soviet city. In Chapter Four Denis Shaw examines how economic restructuring is changing this. At the narrow scale, the 'law on enterprises' and the decentralization of decision-making are changing the role of local government in the Soviet city. At the broader scale, new national investment priorities may alter the nature of the urban hierarchy as old industrial centres decline and new centres of development appear. Economic reform is not, however, confined to urban-based industrial sectors of the economy: agriculture is also experiencing radical transformation. In Chapter Five Judith Pallot examines the changes perestroyka is bringing to the Soviet countryside. The very basis of the farming system is changing, with greater control being placed in the hands of the farmer; after decades of collectivization and socialization, the family is to re-emerge as the organizing unit of agricultural activity. Increased agricultural production is vital to the prospects of economic reform, but can Gorbachev improve living standards in the countryside at the same time as restructuring the industrial base? Stalin's industrial revolution exploited the countryside; might Gorbachev create a new coalition between town and country as symbolized by the hammer and sickle?

One of the cornerstones of economic reform is the drive to promote greater resource efficiency and conservation; increased energy efficiency, for example, can reduce the need to develop high-cost energy resources in remote areas. In Chapter Six Gary Hausladen adopts an historical approach to the relationship between the core regions of the USSR and the peripheral resource-frontier, and discusses the extent to which a new economic reality has reduced enthusiasm for resource projects in Siberia and the Far East. In the past the desire to ensure self-sufficiency and the ideological and strategic benefits of remote area development outweighed the economic costs; will this continue to be the case? Chapter Seven presents a more specific case study. Brenton Barr looks at the relationship between economic efficiency and environmental management in the Soviet forestry industry. The plight of the forest has come to symbolize the international environmental crisis. Can increased public concern for the environment be reconciled with calls for greater economic efficiency and an acceleration of growth rates? Glasnost is revealing the extent of the environmental degradation caused by the Stalinist extensive development strategy and its attitude towards natural resources. Are the economics of market reform compatible with an increased public concern for environmental conservation?

If it is the case that Perestroyka is changing the geography of economic activity and thus regional specializations, then the transportation system must respond with a spatial and functional reorientation. Because transportation was consid ered a 'non-productive' activity under the Stalinist system, it has consistently suffered from a lack of investment. Transportation bottlenecks have been commonplace and a lack of cooperation between the various modes has often aggravated the situation. In Chapter Eight, Robert North examines how the transportation system has responded to perestroyka and whether it will put a brake on reform. The last thematic essay, by Michael Bradshaw, considers the international economic aspects of Perestroyka. During the 1970s Brezhnev used centralized control over foreign trade to inject Western technology into a stagnant domestic economy. Mikhail Gorbachev has acknowledged the need for domestic solutions; nevertheless foreign trade is playing a vital role in economic reform. The Soviet foreign trade system has experienced radical reform and the domestic economy is being opened to greater international competition. New forms of economic cooperation, such as joint ventures, have been introduced. Chapter Nine considers the spatial consequences of internationalization. The final essay in this collection identifies common themes and returns to the issues raised by this introductory essay. What are the geographical dimensions of perestroyka and can the Soviet Union overcome the challenges presented by its geography?

Suggested reading

Bater, J.H. (1989), *The Soviet Scene: A Geographical Perspective*, Edward Arnold, London.

Gregory, P.R. and Stuart, R.C. (1990), *Soviet Economic Structure and Performance*, (4th edn.), Harper & Row, New York.

Lewin, M. (1988), *The Gorbachev Phenomenon: a Historical Interpretation*, Hutchinson Radius, London.

Lydolph, P. (1990), *Geography of the USSR*, (5th edn.), Misty Valley Publishing, Elkhart Lake, W.I.

Massey, D. (1984), *Spatial Divisions of Labour*, Macmillan, London.

Pallot, J. and Shaw, D.J.B. (1981), *Planning in the Soviet Union*, Croom Helm, London.

Schiffer, J.R. (1989), *Soviet Regional Economic Policy: the East-West Debate over Pacific Siberian Development*, Macmillan, London.

Smith, G. (1989), *Planned Development in the Socialist World*, Cambridge University Press, Cambridge.

References

Aganbegyan, A.G. (1988), *The Challenge: The Economics of Perestroika,* Hutchinson, London.

Bandman, M.K. and Linge, G.J.R. (1991), 'Siberian Territorial Production Complexes and Soviet Economic Planning', in Rich, D.C. and Linge, G.J.R. (eds), *The State and the Spatial Management of Industrial Change,* Routledge, London, pp. 40–57.

Blackwell, W.L. (1982), *The Industrialization of Russia,* (2nd edn.), Harlan Davidson Inc., Illinois.

Blackwell, W.L. (1983), 'The historical geography of industry in Russia during the nineteenth century', in Bater, J.H. and French, R.A. (eds) *Studies in Russian Historical Geography,* Vol. 2, Academic Press, London, pp. 387–422.

Bradshaw, M.J. (1990), 'New Regional Geography, Foreign-area studies and Perestroyka', *Area* 22 (4): pp. 315–22.

Cooke, P. (1989), *Localities,* Unwin Hyman, London.

Davis, H. and Scase, R. (1985), *Western Capitalism and State Socialism,* Blackwell, Oxford.

Dibb, P. (1987), *The Soviet Union: the Incomplete Superpower* (2nd edn.), Macmillan, London.

Dienes, L. (1987), *Soviet Asia: Economic Development and National Policy Choices,* Westview Press, Boulder.

Goldman, M.I. (1987), *Gorbachev's Challenge: Economic Reform in the Age of High Technology,* W.W. Norton, New York.

Granberg, A.G. (1986), 'Structural changes and intensification in Siberian industry', *Problems of Economics,* 29 (3): pp. 39–60.

Kingkade, W.W. (1987), 'Demographic trends in the Soviet Union', in *Gorbachev's Economic Plans, Vol. 1,* Joint Economic Committee, Congress of the United States, Government Printing Office, Washington, D.C., pp. 166–86.

Lane, D. (1989), 'Social change, division and control in the USSR', in Scase, R. (ed.) *Industrial Societies: Crisis and Division in Western Capitalism and State Socialism,* Unwin Hyman, London, pp. 168–92.

Lewin, M. (1988), *The Gorbachev Phenomenon: a Historical Interpretation,* Hutchinson Radius, London.

Lovering, J. (1989), 'The restructuring debate', in Peet, R. and Thrift, N. (eds) *New Models in Geography, Vol. 1,* Unwin Hyman, London, pp. 198–223.

McAuley, A. (1985), 'Soviet development policy in Central Asia', in Classen, R. (ed.) *Soviet Interests in the Third World,* Sage, London, pp. 297–318.

Massey, D. (1984), *Spatial Divisions of Labour,* Macmillan, London.

Massey, D. and Meagan, R. (1982), *The Anatomy of Job Loss,* Methuen, London.

Nove, A. (1982), *An Economic History of the USSR,* Penguin Books, Harmondsworth.

Pallot, J. and Shaw, D.J.B. (1981), *Planning in the Soviet Union,* Croom Helm, London.

Sagers, M.J. and Shabad, T. (1990), *The Chemical Industry in the USSR: An Economic Geography,* ACS & Westview Press, Boulder, Colorado.

Schiffer, J.R. (1989), *Soviet Regional Economic Policy: the East-West Debate over Pacific Siberian Development,* Macmillan, London.

Shabad, T. (1969), *Basic Industrial Resources of the USSR,* Columbia University Press, New York.

Storper, R. and Walker, R. (1989), *The Capitalist Imperative: Territory, Technology and Industrial Growth,* Basil Blackwell, Oxford.

Warde, A. (1985), 'Spatial change, politics, and the division of labour', in Gregory, D. and Urry, J. (eds) *Social Relations and Spatial Structures,* Macmillan, London, pp. 190–212.

2

Spatial inequality under Gorbachev

*Ronald D. Liebowitz**

Introduction

Since the founding of the Soviet Union in 1922, Soviet leaders have pledged as one of their major goals the equalization of socioeconomic conditions across regions, between rural and urban areas, and among the state's 100-plus nationalities. While the problems associated with inequality are a concern to all governments, in the Soviet context, the concern is more relevant because of the state's ideology, within which equality represents the hallmark of the socialist system. More concretely, the question of equality touches on domestic and so-called 'ethnic' politics due to the regional specific settlement patterns of many of the Soviet Union's 100-plus ethnic groups. Not only has socialism been touted as the most egalitarian of social systems, but the elimination of inequality was seen as a necessary precondition for establishing communism, and the successful attainment of social and economic equality represented a large part of the state's role in its seventy-year social contract with the Soviet population. The goal of this chapter is to provide an overview of the trends in inequality in the USSR from a spatial perspective over the past three decades, and to evaluate the performance of the Soviet system in reducing inequalities across the union republics and among Soviet nations.

For most Soviet ethnic groups, the equalization of socioeconomic conditions, which was to be accomplished through state-sponsored economic development programs, was supposed to begin the process of eliminating 'national' distinctions and transferring allegiances from individual nations to a common class-based identity. The end result of this process was to be a single unified people, originally referred to as 'Soviet Man' (Connor, 1984; Nahaylo and Swoboda, 1990), whose creation would signal the ultimate victory of communism. Needless to say, this lofty goal was never achieved, nor was the wide acceptance of a class-based 'Soviet Nation' that could replace the stronger primordial attachment of the ethnic group. Nevertheless, the ideals embodied within socialist ideology became the cornerstones of many Soviet social and economic policies, and were used cunningly to differentiate the Soviet system from the exploitative capitalist system.

Official statements on the commitment to develop equal productive capacities among regions and comparable social conditions for all Soviet nations could be found in virtually all Party Conference documents relating to Party and government economic and social programs from Khrushchev's time to the 1970s. A major break in the Party's view of these goals vis-à-vis the general needs of the national economy came in 1972. General Secretary Leonid Brezhnev announced that the 'nationalities' question had been solved completely and irrefutably, which was to imply that the 'official' cause of national differences – social and economic inequality – had been addressed and resolved. The importance of such a statement was not that Brezhnev lied to a quarter billion Soviet people, or that several members of the Politburo may have believed Brezhnev's assessment of the nationalities' question. What was more important in terms of regional economic policy was that Brezhnev was signalling a change in the Party position on economic development priorities. Since equalization in some rationalized way had been attained, there was no longer a need for Soviet development policy to be circumscribed, or at least *as* circumscribed, by what are now seen as irrational economic considerations, made in an attempt to promote equality. Instead, development decisions could be based more on 'rational' economic criteria than on less-than-optimal (economically speaking) egalitarian principles.[1]

Following the 1972 Brezhnev declaration, references to equality and statements promoting equalization became less frequent in speeches given by Party and government officials, and since 1987, official Soviet pronouncements have become virtually free of such discussions, except for recent concern over the effects of a transition to a market economy (e.g. increasing income gaps, unemployment and poverty). Instead, speeches by high-ranking officials, including the General Secretary, have tried to motivate the Soviet population to work harder, take greater responsibility in the workplace, and accept the terms and alleged

*This work was supported by funds provided by the National Research Council for Soviet and East European Research, which, however, is not responsible for the contents or findings.

benefits of a fledgling meritocracy. Attitudes, however, do not change overnight, and much of the public resistance to perestroyka, or what in effect is a new set of terms for the population's implicit social contract,[2] is a function of a seemingly unbending collective psyche that has been moulded and ingrained by socialist ideology.

For decades, the state attempted to build socialism and reduce inequality through wage levelling, compensatory investment strategies and large subsidies, with apparent limited success, although the extent to which one may claim success or failure in this area depends on how one chooses to define and measure equality (Bahry and Nechemias, 1981). Despite the apparent diminution of socialist ideology under General Secretary Gorbachev, most researchers have found that the equality-oriented principles of socialist ideology played a major role in the rhetoric – if not in the actual decision-making – of major economic development decisions (Dyker, 1983; Bahry, 1987; Wagener, 1973). But while there are clear examples where Soviet policy has been very effective, such as in reducing illiteracy, increasing educational attainment levels in both urban and rural areas and providing universal basic health care to all Soviet citizens, the system has not been successful in many other areas, as the empirical section of this chapter will show.

Throughout the 1950s, and until the early 1970s, Soviet economists and economic geographers wrote often of 'proportional development' (Holubnychy, 1973; Woroniak, 1973; Zakumbayev, 1977; Nove, 1978), the 'locating of productive forces so as to insure the equal production and distribution of goods and services', plus other socialist-oriented schemes whose blueprints were long on goals but short on specifics. There were countless calls advocating 'regional self-sufficiency', while there were at least as many promoting specialization for 'greater efficiency within the socialist economic division of labour'. Needless to say, a viable Soviet location theory – at least one that could yield viable development schemes for equalizing regions – never emerged (Huzinec, 1977), and few prescriptions were put forth that could help mitigate the inconsistencies between stated economic priorities and the realities of the country's economic geography. Instead, broad goals calling for *both* economic *efficiency* and regional economic *equality* were set; but efficiency in the free-market sense remained difficult to estimate (Dyker, 1983) and equality meant different things to different interests at different times (Bahry and Nechemias, 1981).

Despite a significant literature on the question of inequality in the USSR, several problems intervene to make it extremely difficult to arrive at meaningful conclusions on the level of, and trends in, inequality across the Soviet Union.

Measuring inequality

The difficulties in estimating efficiency or any economic indicator that could then be used to evaluate inequality begin with the absence of a rational pricing system. The lack of such a system prevents the wide range of estimates necessary to make rational allocation decisions, and hinders regions or republics from making choices between producing for local and domestic markets versus producing for international markets (Brada, 1990; Gardner, 1990). At present, data used to estimate comparative wealth and inequality are based on prices that are determined largely arbitrarily and therefore reflect the whims of Soviet planners and vast subsidies rather than a region's actual material wealth or its value of production. The conflict between pursuing efficiency versus equality is not simply a matter of the ideological bearings of the Soviet leadership at any one time. Instead, the tension between initiating a development policy that addresses the goals of equalization or maximum economic growth and hence efficiency is rooted in the structure and contradictions of the Soviet planning system. Specifically, competition between the ministry-dominated branch-sectoral planning approach and the more equality oriented territorial-based (i.e. regional) planning approach leaves the issue of efficiency versus equality unclear at best. Since the 1950s, except for a brief period during Khrushchev's *sovnarkhozy* decentralization attempts, the branch-sectoral approach to planning, in which large ministries make investment and development decisions based on their calculations of efficiency and profit potential, has dominated most major aspects of Soviet economic planning (Schiffer, 1989). This dominance has extended to all major industrial and infrastructural projects since the Second World War and continues to be a major impediment to significant economic reform. Due to the branch-sectoral approach to planning, the performance of Soviet regional development policies, in so far as its commitment to reducing regional and socioeconomic inequality is concerned, has been difficult to isolate and assess. That is, if one accepts the dominance of the sectoral-branch approach to planning, changes in equality or inequality must be seen to a large degree as an unintended consequence – almost as a residual – of ministerial planning. Although it is clear this method of planning and allocating investments directed Soviet planning over the past three decades, it is not as clear how strongly regional considerations, which include nationality and regional concerns, entered in the decision-making process among those in the higher echelons of the ministries in Moscow.[3]

Despite the dominance of ministerial planning over territorial-based planning, which would *a priori* place the goals of equalization behind those of overall economic growth, many Western researchers (Jones and Grupp, 1982; Bahry, 1987; Nechemias, 1980; Liebowitz, 1987; Schroeder, 1973 and 1981; McAuley, 1979; Koropeckyj, 1972; Spechler, 1979; Clayton, 1975; and others) have tried to evaluate the effectiveness of Soviet social and economic policies in reducing inequality across regions and/or ethnic groups.[4] The results of these works are mixed, depending on whether the *intent* of the policy or its *effectiveness* in reducing inequality was the focus of the study, and on which geographic scale of analysis the study was performed. Unfortunately, relevant data on a more detailed scale than the union republic has been hard to obtain, so most work has been done on the general republican scale of analysis.[5] Works of this sort by Soviet academics have been rare due to the (formerly) sensitive and overtly political nature of issues associated with inequality. The small number of Soviet-authored empirical studies that have addressed regional inequality (Rutgayzer, 1974; Mil'ner, 1969 and 1980; Mil'ner and Gilinskaya, 1975; Zakumbayev, 1977) were all done on a very gross scale of analysis (using either quadrant or macro-regions), and therefore could yield only limited information about social and economic inequality in a state as large and diverse as the Soviet Union.

Needless to say, studies of social and economic inequality in general, and their spatial dimensions in particular, have been lacking, due largely to data constraints in the West and ideological constraints in the Soviet Union. Published data are often irregular, hindering time-series analysis, and coverage among territorial units often differs in definition and quality (consistency) from one period to another. And there are other factors that have made it more difficult to assess the success of the equalization goals of the Soviet Government, including the simple issue of defining relevant terms. Should 'equality' mean the equal attainment of specific so-called objective social and economic indicators for an entire population? Should equality be measured by comparing 'average' scores of selected social and economic indicators for ethnic groups or for the populations of what are arguably arbitrary administrative units? For analyses based on areal units, what scale of analysis is most appropriate for measuring spatial inequality in a country as large and diverse as the Soviet Union? And, how does one standardize these social and economic data when their reliability is questionable – ironically, even more so since glasnost.[6]

While those familiar with Soviet data in the preglasnost days felt they had developed a healthy and necessary scepticism about official Soviet data, the avalanche of new information is difficult to evaluate and use because Soviet academics themselves have begun to question the veracity of official statistics published both before *and* after glasnost (Selyunin and Khanin, 1987). Some charge that before glasnost, data were manipulated so as to present a better picture than that which actually existed; after glasnost, some claim that data from specific periods, most notably corresponding to the late Brezhnev era or the *zastoy* (stagnation), have been manipulated to make the period look worse than it was in order to highlight the need for reform, and to rally support for Gorbachev's reform program. Consequently, caution and careful data evaluation must accompany the use of the new statistical information now available for studies on Soviet society.

In this chapter, I shall use some of the newly released data on Soviet society to provide an overview of social and economic conditions throughout the Soviet Union. Specifically, I will assess the long-time Soviet commitment to improving the socioeconomic conditions of the Soviet Union's ethnically diverse and geographically dispersed population, and to reducing inequalities among the population and regions. I shall highlight the trends in selected socioeconomic conditions over time,[7] and then place these and possible future trends within the context of the Gorbachev reform program. This last task will assess how perestroyka in its broad conception is likely to affect trends in inequality among the state's fifteen union republics.

Aside from providing new information on the level of inequality among Soviet regions and how inequality has changed over time, this chapter will also evaluate the record of the Soviet highly centralized allocative mechanism in allocating increasingly scarce resources among disparate territories. The track record of resource allocation in a socialist state carries great political weight because unlike capitalist states, in which the private sector decides to a large degree where investment resources will go, in the (centralized) socialist state, the government has virtually complete control over the direction and level of resource allocations. As a result, disenchanted groups in socialist states have an identifiable target at which to lodge protests – the central government – if they feel they are receiving less than their fair share of state investments.

Data and methodological questions

Since glasnost was introduced by President Gorbachev in 1985, and especially since 1987, the availability of social and economic data on Soviet society has changed dramatically. The newly published data should serve to breath life into what had become a moribund activity among Sovietologists – data-based research on the Soviet Union and its constituent republics and nations – and will force researchers today to rely less on conjecture when formulating research questions on Soviet society than in the past. Questions about comparative living conditions across regions, or public attitudes on previously sensitive political issues should soon be addressed in greater detail than ever before.

In this chapter, I shall assess social and economic inequality in the Soviet Union from a spatial perspective in the following way. I shall use descriptive statistics for selected social and economic indicators for the state's fifteen union republics.[8] This first step provides a summary of various social and economic conditions, and in and of itself provides an overview of inequality in the Soviet Union. I shall use both weighted and unweighted coefficients of variation (CV)[9] plus quintile ratios (QR).[10] The weighted measures take into account the population size differences among the fifteen republics. With the weighted measures, scores for the Russian Republic, for example, would count for more than 1/15 the weight they would with unweighted measures, since each score is multiplied by the population share of its respective republic. Instead of the Russian Republic contributing 6.6 per cent (1/15) of the all-Union (CV) measure (as it would in the unweighted example), for 1989, it would account for 51.4 per cent of the overall Soviet measure.

Both the weighted and unweighted CVs are useful statistics with which to measure inequality, but they answer different questions. Geographers would more readily use the unweighted measures than other social scientists because of their interest in location as a function for studying variation and because of their emphasis on *spatial* comparisons. For example, in the case of estimating inequality among the Soviet republics, a geographer might ask how differences in consumption, investment, income or any other variable vary depending on location. Such a question is less concerned with the absolute level of inequality among the Soviet population than it is with differences *among republics*, because the emphasis is on differences according to locale. Economists and sociologists, on the other hand, would be more apt to use weighted measures, since their concerns are more with the overall level of inequality within a population or in an entire political entity, and their research questions would concentrate on inequality with less regard for the specific locational dimension. In short, the weighted measure assesses differences *within a population,* while the unweighted measures estimate differences *among territorial units,* in this case, the fifteen Soviet republics. The results and significance of both the weighted and unweighted tests will be discussed in the paper.

As part of the analysis, I shall look at the issue of inequality over time in order to assess how well the Soviet system performed in reducing inequalities. The state's degree of success in this area is important, in the broad context, because the legitimacy of any state/government depends on how well problems such as poverty and inequality are addressed. In the Soviet example, the success in dealing with the question of equality is of great importance since one of socialism's self-proclaimed strengths, and one that has been used to justify some of the Soviet system's shortcomings, is that socialism is the system best equipped to reduce social and economic inequality and foster prosperity. While this may now be seen as a moot point since the Soviet Union appears to be moving away from its socialist bearing,[11] it remains important in so far as our understanding of how central control of economic and political decision-making (i.e. decisions on industrial location and economic development, etc.) relates to trends in inequality and to maintaining political stability in states as geographically and ethnically diverse as the Soviet Union. Finally, the last part of this analysis looks specifically at the state's *commitment* to solve some of the measured inequalities by analysing investment flows and budgetary trends since the 1950s. Because the levels and direction of capital allocations associated with large-scale regional economic development have, to date, been controlled by the state and specifically by central ministries in Moscow, analysing the trends in regional investment flows provides a good reflection of Soviet development objectives and priorities.

All of the data used in this analysis were published in the Soviet Union. Most of the data were first published in press bulletins by Goskomstat (State Committee on Statistics) and later appeared in an assortment of Soviet publications.[12] Where possible, figures were cross-checked with second and third sources for consistency and plausibility, but because many of the data have only recently been published, confirmation for many variables was impossible.

Social and economic inequality in the Soviet Union

A republican-level analysis of the data (see Table 2.1) confirms many similar patterns found in most market economies; there is great variation in most social and economic indicators across regions and between urban and rural areas, and most of the gaps are widening. In addition, trends in state-controlled investment allocations also show growing divergence between the state's more developed republics (MDRs) and less developed republics (LDRs). While these trends may suggest an overall failure of the Soviet system to meet its goals of 'equalizing' regions and the standards of living of Soviet nations, there are some caveats to this conclusion that need to be addressed at the end of this section.

Trends in *per capita retail trade* data,[13] an often-used surrogate measure for personal consumption, reflect the large variation in material conditions among republics and between rural and urban areas. By 1988, a distinct regional hierarchy, one that was far less apparent in 1960, was in place, with the Baltics and the four European republics ranking highest, and the Central Asian and Transcaucasian republics far behind.[14] These data exclude goods and services obtained from the 'unofficial economy', which is significant,[15] especially in the agricultural areas, but the *trends* for retail turnover in the state sector are important. Retail trade per capita ranges from between 26 to 53 per cent above the all-Union average in the Baltics to as little as half the all-Union norm or one-third the Estonian level in Tadzhikistan. The difference between urban and rural retail trade figures reflects similar inequalities, but it should be reiterated that the unofficial sector adds, on average, more to rural residents than to those living in urban areas. In 1988, per capita retail trade was 123 per cent higher in urban areas than in rural areas for the entire country, and in some republics, the differences were greater than 200 per cent. And the variation in rural areas across republics is quite a bit more than among urban areas: the QR for urban areas in 1988 was 1.87 while for rural areas it was 2.61.

More important than these relationships, perhaps, is the apparent *increase* in the variation in per capita retail trade among republican populations and within the entire Soviet population over the past three decades. In 1960, the unweighted CV for per capita retail trade was 26.8 per cent; in 1988 the index increased to 32.9 per cent, an

Table 2.1 *Trends in selected social and economic indicators*

		1980	1985	1988
National income produced*	CV (unweighted)	25.7	27.4	27.6
	CV (weighted)	14.5	19.4	23.6
	Quintile Ratio	1.49	1.96	2.15
		1960	1980	1988
Wages	CV (unweighted)	8.9	7.9	9.4
	CV (weighted)	6.4	5.7	7.4
	Quintile Ratio	1.3	1.2	1.2
		1970	1980	1988
Infant mortality	CV (unweighted)	39.2	47.8	53.4
	CV (weighted)	23.5	35.9	39.7
	Quintile Ratio	2.58	3.48	3.77
		1971–75	1981–85	1987
Agricultural production	CV (unweighted)	35.6	32.5	36.9
	CV (weighted)	22.1	22.1	24.8
	Quintile Ratio	2.78	2.11	2.65
		1971–75	1981–85	1987
Housing outlays*	CV (unweighted)	29.4	28.9	28.1
	CV (weighted)	22.7	28.4	33.6
	Quintile Ratio	2.06	2.33	2.14
		1960	1988	
Retail trade*	CV (unweighted)	26.8	32.9	
	CV (weighted)	15.5	18.2	
	Quintile Ratio	2.06	2.44	
		1960	1987	
Students in higher ed. (standardized)	CV (unweighted)	21.6	12.6	
	CV (weighted)	16.1	9.1	
	Quintile Ratio	1.83	1.33	
		1980	1988	
Savings accounts*	CV (unweighted)	46.3	51.5	
	CV (weighted)	26.4	30.7	
	Quintile Ratio	3.6	4.54	
		1980	1987	
Living space*	CV (unweighted)	22.3	24.1	
	CV (weighted)	12.4	15.1	
	Quintile Ratio	1.76	1.83	

		1956–60	1961–5	1966–70	1971–5	1976–80	1981–5	1986–8
Capital investment*	CV (Un)	31.8	29.1	21.2	23.6	23.8	24.4	28.6
	CV (W)	21.7	13.7	14.5	17.5	21.3	22.6	26.2
	QR	2.04	1.86	1.69	1.85	1.89	1.96	2.22

*Based on per capita measures

Sources: Goskomstat bulletins nos. 1–7; 9–12, 1989 and number 16, 1990; *Statisticheskie materialy ob ekonomicheskom i sotsial'nom razvitii soyuznykh i avtonomnykh respublik, avtonomykh oblastey i okrugov:* chast' 1 i chast' 2; *Narodnoye Khozyaystvo SSSR v 1987g, and 1988g; Trud v SSSR; Naseleniye SSSR 1987, 1988; Promyshlennost' SSSR; Kapital'noye Stroitel'stvo SSSR;* and *Sel'skoye Khozyaystvo SSSR.*

increase of about 23 per cent. The weighted CV showed a similar trend, increasing 14.8 per cent. Similarly, the QR increased from 2.06 to 2.44 between 1960 and 1988, providing more evidence of the growing gap between the LDRs and MDRs.

Differences in consumption (or its surrogate, *per capita retail trade*) and social conditions in general are tied closely to differences in regional economies. Part of the long-standing Soviet goal to reduce inequality was to develop the national economy so that the productive capacities of the state's economic regions would be 'more equal' and 'self-sufficient' while still contributing to the national economy. The non-uniform distribution of labour and natural resources, however, intercedes in the Soviet Union as it does elsewhere so that a far more skewed pattern of economic development than that which Soviet planners desired is currently in place across Soviet territory. Since the Second World War, while impressive development has taken place in many of the state's formerly relatively backward areas, the concentration of economic production in the western parts of the state has not diminished (see Chapter six). That is, the older and more established industrial regions of the state continue to produce (per capita) more than the southern tier, and their dominance has increased.

Trends in the relative share of *national income produced* in each republic reflect the increasing concentration of economic production during the past decade. The unweighted CV of per capita national income produced (by value) increased only modestly (7.3 per cent), but the weighted measure, which takes into account the dominance of the Russian Republic (RSFSR) and Ukraine, increased 63 per cent, and the QR increased 44 per cent.[16] Ratios of per capita national income produced (in roubles) for the Russian Republic on the one hand, and Uzbekistan, Kirgiziya and Tadzhikistan on the other were 1.82, 1.94, 2.01 in 1980, and increased to 2.34, 2.28 and 2.68 by 1988. Differential rates of natural population increase among the republics can account for some of the decline of the LDRs vis-à-vis the MDRs,[17] but the increasing dominance of the developed Slavic republics is consistent with the de- emphasis of the LDRs in Soviet investment policies since the early 1970s, particularly for productive investments (discussed below). In 1988, the RSFSR, Ukraine and Belorussia produced 81.5 per cent of total national income (up from 79.9 per cent in 1985 and 80.0 per cent in 1980) with 72.6 per cent of the state's population, while the four Central Asian republics accounted for 5.6 per cent of the national income produced (down from 6.24 per cent in 1985 and 6.19 in 1980) with slightly more than 12 per cent of the Soviet population. It is likely that these figures would be even more skewed if Soviet output (resources and machinery) were valued at world prices.

Agricultural production has remained concentrated over the past fifteen years, and per capita measures among the republics have become more variable. Since 1971, the MDRs (the three Slavic republics and the Baltics) have retained about 79 per cent of the value of agricultural production despite employing a significantly smaller share of their labour forces in agriculture than the southern tier republics. The four Central Asian republics increased their share of the state's total agricultural production from 7.4 to 8.1 per cent of the Soviet total, but during this period their share of the Soviet total population increased from 8.3 per cent to slightly more than 12 per cent. More significant, perhaps, is that upwards of one-third of the Central Asian labour force is currently employed in agriculture compared to fewer than one-eighth in the Russian republic and one-sixth in the Baltics (USSRh, 1988, pp.16–17). The combination of these trends has caused the variation in per capita agricultural production among the fifteen republics to have increased (12.2 per cent), especially during the past decade.[18]

Industrial wages have increased steadily throughout the Soviet Union since the late 1950s and, unlike other social and economic indicators, offer little evidence of large increases in inequality among regions. Although there is a predictable regional hierarchy to average wages, the differences are not as great as they are for other social and economic indicators. These favourable trends are not surprising: wage policies are relatively easy to administer in a command economy, and wage levelling across regions has been a stated policy for years (McAuley, 1979). Where large wage differences do exist, they are mostly a function of the need to attract labour to less-than-hospitable regions, such as to West Siberia, East Siberia and the Far East (Mil'ner, 1979; Rybakovskii, 1969;

Topilin, 1975; Serdit, 1969). The CVs, both weighted and unweighted, declined nominally between 1960 and 1980 and then increased marginally between 1980 and 1988 to levels slightly greater than they were in 1960.

The size of personal *savings accounts* is another economic variable that varies by region, and the differences have increased over the past decade. Despite the fact that there are more goods to purchase, people in the Baltic republics, the Ukraine and in Belorussia maintain average savings accounts up to five times those of the populations in the southern tier (1647 roubles in Lithuania compared to 280 roubles in Tadzhikistan). Since 1980, savings have increased on average 75 per cent (about 9 per cent/year) due in large part to rising wages and great suppressed consumer demand. This combination has heightened concern over inflation, which also differs regionally. But while (per capita) savings have increased on average at least 45 per cent in every republic, both the weighted and unweighted CVs increased significantly over the past decade, as did the QR (from 3.62 to 4.54), reflecting greater divergence or inequality in this contributor to personal wealth. The amount of one's savings, like other variables in this kind of analysis, is subject to a host of varying cultural- and regional-specific characteristics unrelated to a region's economy; its inclusion here is intended to add to the larger picture of inequality rather than to create its own.

If the preceding economic measures describe a regionally stratified society, most of the social and demographic data available do more of the same. *Infant mortality rates* (IMRs), often viewed as the best single indicator of a population's quality of health care, vary considerably among Soviet republics, and the variation has not declined during the past twenty years. IMRs are five times higher in predominantly rural Turkmenistan than they are in the Baltics, and the CVs for the fifteen republics have increased from 23.5 per cent (weighted) and 39.7 per cent (unweighted) in 1970 to 39.2 (weighted) and 53.4 (unweighted) per cent in 1988. Similarly, the QR increased from 2.58 in 1970 to 3.77 in 1988, reflecting the increasing gap between the MDRs and the republics of the southern tier. While one must exercise caution in evaluating demographic data like IMRs,[19] such great variation in an indicator like infant mortality reflects the great differences in pre-natal and general medical care, as well differences in the overall social environment across the Soviet Union. During the past five years, several Soviet newspapers have published in-depth articles linking environmental degradation and increases in mortality, and especially infant deaths. Water contamination from pesticides and fertilizers, particularly in areas of extremely high infant mortality, is reported to be responsible for exceptionally high incidences of infant deaths caused by infectious diseases. There is a large variation in deaths caused by infectious diseases across regions, with rates 20–25 times higher in Central Asia than in Estonia and Lithuania (USSRe, 1989, pp. 680–4). Almost 37 per cent of infant deaths in Tadzhikistan in 1986 were due to infectious diseases, while in Estonia, the RSFSR and Moldavia the corresponding percentages were 6.7, 10.8 and 14.4 per cent (USSRe, 1988, pp. 400–7).

A second demographic variable, *life expectancy*, further reflects the north-south dichotomy in living conditions, with the populations of the Baltics and the other European republics living on average four to eight years longer than those living in Soviet Central Asia.[20] While life expectancy throughout the country has begun to increase following an unprecedented decline in the late 1970s and early 1980s, the gap among republics has not narrowed. The differences in life expectancy between the MDRs and LDRs are especially large when one considers the age structures of their respective populations; because death is an age-specific variable, younger populations, like those in the southern tier, have lower aggregate death rates than older populations.[21] That the younger populations in the Soviet Union (those of the LDRs) have shorter life expectancies reflects the very high death rates among infants and the elderly, together with substantial differences in health-care delivery systems across republics.

An obvious issue related to the state of health conditions and their variation across regions is the amount of financial resources given to health care and the availability of medical facilities throughout the country. For the last five-year period for which data were available (1981–5) and the following three years (1986–8), the range of *resources allocated for health* is, not surprisingly, quite large, and the ranking of the republics largely conforms to the already familiar MDR–LDR hierarchy: the Baltic and European republics spend the most on health care, allocating, on average, 42 per cent more (roubles per capita) than the southern tier republics, and 47 per cent more than the four Central Asian republics (USSRa, 1987) for the 1981–5 period. The differences for the 1986–8 period were 37 and 41 per cent (USSRa, 1988 and 1989). Because allocations for health care have become primarily the responsibility of the individual republics,[22] some have argued that the centralized budgetary system should not be blamed for such great regional differences in allocations, which is a curious argument. If the central budget cannot assist the poorer republics with the increasing costs of medical care and other social needs, the system loses one of its most appealing characteristics and explains, in part, the strong desire for greater republican autonomy now being voiced by even the poorest republics. It is one thing to have little control over large industrial projects that are part of a larger 'national plan' if there are associated benefits. It is quite another issue to give up control over one's own territory, receive little in return, and be asked to finance a greater proportion of a rapidly growing social services bill, as seems to be the situation in most of the poorer republics. Conversely, the wealthier republics may see themselves as finally being in the position to cast off the economic burden of the LDRs.

Quantitative indices for *medical facilities and personnel* have long been used by Soviet authorities to publicize the success and superiority of the socialist system in meeting the everyday needs of a modern society. Unfortunately, the number of facilities does not necessarily relate to the quality of care, and the Soviets themselves have begun to criticize the use of quantitative-based indices for measuring social provisions. Problems such as accessibility for many rural inhabitants due to the lack of reliable rural roads and inadequate public transportation are often cited in the Soviet press. During the past two decades, the Soviets have reduced inequalities in the supply of several healthcare-related facilities, including hospital beds, doctors and medical personnel; the CVs and QRs for all three variables either remained constant or declined. While it is impossible to evaluate the *quality* of care from these indicators, the Baltic republics and the RSFSR ranked highest in terms of their quantitative indices for each of these indicators; the southern republics ranked lowest.

Regional inequalities in *educational attainment* have been reduced perhaps more impressively than inequalities in any other social and economic measure for which data are available. According to the Soviet census, illiteracy has been virtually eliminated (USSRb, 1972, pp.570–1) and educational attainment levels have increased impressively throughout the entire union. Between 1960 and 1987, the weighted and unweighted CVs and the QR for the number of graduates with higher education all declined significantly,[23] and the ranking of republics appears random, with Uzbekistan, Azerbaydzhan and Kirgiziya near the top of the list.[24] In contrast, *per capita living space* varies significantly from republic to republic and between urban and rural populations; the gaps have increased over the past decade. Residents in the southern tier have about half the living space of people in the Baltics and Georgia. A large part of the regional variation is due to family size, but the gaps between the MDRs and LDRs are present in both rural and urban areas, and the CVs and QR have increased for both areas during the 1980s. Not surprisingly, the per capita living space in the rural areas of the MDRs is greater than in the MDRs' urban areas, while the relationship is reversed in the LDRs. This difference is largely a function of the uniformly large family size in the rural areas of the LDRs and the high percentage of non-indigenes – mostly Russians – living in urban areas of the LDRs, while birth rates even in rural areas of the MDRs have fallen considerably.

The variation in living space among republics is partly due to noticeable differences in per capita state and private *expenditures on housing*. The European regions of the state spend upwards of twice the amount (per capita) on new housing as the southern tier – more than 110 roubles in Lithuania and the RSFSR to less than 50 roubles in Azerbaydzhan and Tadzhikistan for the 1981–5 period – and while the unweighted CVs for this variable have remained stable, the QR has increased since 1980 (2.06 to 2.33). The seemingly conflicting results of these two measures reflect how the gap between absolute highest and lowest measures (republics) could increase significantly while the overall variation among all fifteen measures (republics) may not. Throughout most Soviet cities there remain long waiting lists for apartments, so differences in spending on housing must be viewed with some caution. Large housing expenditures in the western portions of the state are still partially a function of the lingering effects of the Second World War, when so much of the housing stock in the European part of the country was destroyed; but even in the south, due to the region's demographics, the demand for housing is well beyond supply. Unfortunately, the state of the national and local economies will preclude large infusions of state capital for new housing.

The inequality in new housing stock among regions is magnified by the role industrial ministries play in the housing market. Ministries own and build a large number of flats for their workers, so the populations of those areas in which the industrial ministries play a dominant role have considerably more funds on which to rely than state and private investment funds. Since industrial ministries play a far greater role in the economies of the MDRs than in the southern tier, the real gap between regions is even larger than the per capita expenditures suggest.

Differences in living standards among rural areas, and particularly among collective farm households, are as pronounced as, and are perhaps greater than, overall inter-republican differences. A press bulletin released by Goskomstat (Press Bulletin No.5, 1988) reported the percentage of collective farm households with an assortment of *household amenities* (by republic); while the differences in these measures among republics are substantial, the absolute level of provisions (or lack thereof) union-wide is startling. Fewer than a quarter of collective farm households in all four Central Asian republics and Kazakhstan have running water; fewer than a quarter of the households in all the republics except Belorussia, the Baltics, Moldavia and Armenia have indoor bathrooms; sewerage is almost non-existent outside the Baltics, Armenia and Moldavia; and hot water is available in less than fifteen per cent of collective farm households in every republic except the Baltics and Armenia. Clearly, despite the rhetoric about equalization, and whatever the intent of Soviet planners, large gaps in living standards persist both among republics and between urban and rural areas.

Governmental inputs: capital investments

Since collectivization and the start of the Five-Year plans, the Soviet system of resource allocation has been strongly centralized. Consequently, investment allocations to a great extent reflect the state's agenda in terms of economic and political priorities. The desire to 'equalize' regions has always had to compete with the need to promote maximum economic growth (equality versus efficiency). By analysing trends in regional investment allocations since Khrushchev, one can gain a sense of the state's commitment to equalization, or at the least, how 'equal' investment allocations have been since the 1950s.[25] By combining these findings with the findings on the trends in equality, we can come to some conclusions on the effectiveness of Soviet development strategy in reducing inequality.

Per capita investment allocations for the three decades preceding the Gorbachev era can be evaluated in many ways. Aggregate per capita figures support the contention that the Russian Republic receives favourable treatment, while the southern republics, except for natural gas-rich Turkmenistan, receive far below the all-Union mean. These conclusions, however, must be qualified. The Russian Republic measure is a bit distorted, because its superior position is only recent, and is largely due to the requirements of the republic's energy and resource-extraction industries. Diminishing energy production in most Soviet regions during the 1960s and the difficult geography of Tyumen's oil and gas exploitation, combined to force huge investments into West Siberian energy development during the past few decades: a development that continues to distort the spatial profile of investment allocations. In fact, the development of Tyumen's oil and gas reserves, and further work on the Baykal-Amur railroad, accounted for well over half the growth in investment in the Russian Republic since the mid-1970s (Dienes, 1989). The distorting weight of a few large and sparsely populated oblasts, which gives the RSFSR its apparent favoured standing, shows why the more detailed oblast level of analysis is needed. While investment allocations in Soviet Central Asia were well below the all-Union mean for the entire period, a large part of the region's aggregate capital deficit has been accumulated over the past fifteen to twenty years. Central Asia's declining position is partly due to the region's rapid population growth, but a de-emphasis on 'equalization' in favour of the need to increase the state's return on its investments is also responsible.

The greatest variation in per capita investment allocations occurred during the 1956–60 period; the least variation as measured by both weighted and unweighted CVs and the QR was during the 1966–70 period. Since 1966–70, the CVs for each successive period have increased, reflecting a more directed investment strategy aimed at the developed republics. The increasing QR, which, again, compares the median of the first quintile scores with the median of the last quintile, reflects the growing divergence in capital allocations between the top- and bottom-ranked republics since the 1966–70 period.

Several questions must be asked before one can draw conclusions from these data. How realistic is it to test for 'equality' in the strictest sense of the word in a country as large and diverse as the Soviet Union? Would it be fairer, as some have claimed, to evaluate Soviet investment allocations, especially in the relatively backward regions of the country, against what was likely to have occurred had all the regions been on their own and left to generate investment funds from their own economies?[26] Clearly, per capita investments have not become 'equal', and in fact have become less so since 1970. The structure of investment in the poorer republics has not reflected an economic development strategy geared toward diversified development or the kinds of economic activity that could promote greater wealth and reduce inequality. A much larger share of the southern republics' investment has gone and continues to go to agriculture and to 'non- productive' sectors of the economy, while the preponderant share of industrial investment continues to flow into the developed regions.[27] Such investment patterns over the past thirty years have helped to perpetuate the well-established differences in the economic structures of northern and southern republics, which directly contribute to the high degree of social and economic inequality throughout the country (Dienes, 1987; Rumer 1989).

Data on the individual labour forces of the fifteen republics reflect a well-defined division of labour, stratified along regional lines: while over 42 per cent of the labour force in the RSFSR was employed in industry in 1987, the corresponding figures for Uzbekistan, Azerbaydzhan, Tadzhikistan and Turkmeniya were 24, 26, 21, and 21 per cent. Similarly, while only 14 per cent of the RSFSR's labour force was employed in agriculture, the respective figures for the aforementioned LDRs were 38, 34, 42 and 41 per cent. This spatial division of labour also has an ethnic dimension: the percentage of a republic's industrial jobs held by its titular nationality (that is, in its own republic) is far greater in the MDRs than in the southern tier. While Russians hold 83 per cent of industrial jobs in the RSFSR, the figures for Kirgizians in Kirgiziya is 25 per cent, for Tadzhiks in Tadzhikistan it is 48 per cent, for Kazakhs in Kazakhstan it is 21 per cent, and for both the Uzbeks and Turkmenis it is 53 per cent in their respective republics (USSRh, 1988, pp. 22–3). In addition to reflecting great differences in the economic structures of republics, such a division of labour raises the question of how industrial jobs will be filled (and by whom) if and when economic autonomy for the republics becomes a reality – especially if autonomy leads to the kind of large-scale out-migration of non-native, and mostly Russian, industrial workers that has already begun in some southern tier republics.[28]

Given the trends in capital allocations and the evidence that under the highly centralized allocative mechanism capital investments have not become more equal, it is not surprising that gaps in most indicators of social and economic inequality have not narrowed, and in most cases, have widened. The Soviet system, then, appears to have failed to meet its goal of reducing inequality among regions and between urban and rural areas after more than seventy years of socialism. However, some researchers[29] counter criticism of empirical studies that conclude the Soviet system failed to reduce inequality by suggesting that conditions in the relatively backward areas of the Soviet Union would be even worse today if it had not been for Soviet rule and its centralized control of investment resources. This counterfactual argument is difficult to refute. One cannot know with any certainty the path of development individual republics or the collective region might have chosen following the break-up of the Russian

Empire – are Afghanistan and Pakistan the standards against which such an analysis should proceed? Boris Rumer (1989) and others argue that Soviet development policies have had a devastating effect on the Central Asian economy (and its environment) and that Central Asia is little more than an internal colony of the dominant Slavic core. Indeed, Soviets themselves echo this charge quite regularly in the present spirit of glasnost. A member of the USSR Congress of People's Deputies, M. Bronshteyn, in addressing regional tensions due to the siphoning of wealth from some republics, most notably the Russian Republic, recently stated (FBIS–SOV–90–041, 1 March 1990, p.81):

> Who has benefited? Could it be the republics of Central Asia? However, their representatives refer justifiably to the tremendous losses inflicted on their republics due to the single-crop economy imposed by the center and to low cotton prices . . . . Undoubtedly, the Soviet Union as a whole should make up, for example, the loss inflicted on [these] republics as a result of their external orientation toward a single-crop cotton economy while procurement prices were unjustifiably low.

Such an argument, despite its long list of supporters, cannot assume that Central Asia's economy would be more diversified and necessarily more prosperous, or its environment less fouled by the application of 'modern' agricultural techniques and poorly planned industrial development, if the region had not been part of the Soviet Union. The region would probably have traded its dependence on Moscow for a similar economic relationship with the United States and/or West Europe and faced similar economic problems.

It is clear, however, that within the Soviet framework, the relatively backward regions, while not receiving the levels of investment of the more developed republics, have been subsidized to some degree. Economists have argued for decades that the relatively wealthier regions, especially the Ukraine and certain regions of the RSFSR, have financed the development of the more backward regions through transfer payments and other financial mechanisms (Koropeckyj, 1977; Melnyk, 1977; Gillula, 1979), but the absence of a rational pricing system at both the wholesale and retail levels makes it very difficult to estimate inter-republican subsidies.

Ratios of the shares of national income produced relative to the shares of state investments can serve as a crude surrogate measure for state subsidies from the centre to the republics. The ratios for the LDRs and especially for the Central Asian republics (and Kazakhstan) have declined noticeably since the mid-1970s; by the mid-1980s the entire southern tier (except Kirgiziya) ranked in the bottom half of all republics for this measure (see Table 2.2). While the ratios for all the republics declined noticeably, which reflects the decline in the overall productivity of the state's economy, the most surprising ratios are those for the RSFSR and Lithuania, which were both below the national average. The low ratios for these relatively wealthy republics reflect more the irrational Soviet pricing system, which underestimates the resource wealth of the RSFSR and the consumer products' industries of Lithuania than the relative efficiencies of the two republics.

Recently published data on the volume of trade by republics, calculated with both Soviet and world prices, reflect how the Soviet pricing system can alter the picture of inter-republican economic relations (*Vestnik Statistiki,* 1990, p.49). If world prices replaced internal prices for all goods, nine republics – Belorussia, Uzbekistan, Georgia, Azerbaydzhan, Lithuania, Moldavia, Latvia, Armenia and Estonia – would be worse off (*menee blagopriyatny*). Using world prices, five republics would be in about the same position, and only one republic, the RSFSR, would be significantly better off: and the difference would be enormous. The Russian republic would show a 30.8 billion rouble surplus (for 1988) instead of its reported 33.3 billion rouble deficit. This dramatic turn-around is due to the way in which the domestic pricing system undervalues petroleum products, natural gas, ferrous and non-ferrous metals and certain kinds of machinery (*Vestnik Statistiki*, 1990, p.50). Conversely, Soviet internal prices overvalue light industrial products and food products, which explains why the nine republics, all of which have specific agricultural strengths but relatively few natural resources, do better under the Soviet price system. Such an analysis is problematic because intermediate costs of production have not been recalculated and market forces (supply, demand, quality, etc.) were not factored into the world prices used for the Soviet products. But the dramatic difference for the Russian Republic does

Table 2.2 *Ratios of national income produced/state capital investments*

	1976–80		1981–5		1986–8
AZERBAYDZHAN	0.90	ARMENIA	0.89	UKRAINE	0.72
GEORGIA	0.87	GEORGIA	0.88	BELORUSSIA	0.72
ARMENIA	0.86	BELORUSSIA	0.83	LATVIA	0.70
LATVIA	0.83	LATVIA	0.83	GEORGIA	0.70
BELORUSSIA	0.78	UKRAINE	0.83	MOLDAVIA	0.69
ESTONIA	0.74	TADZHIKISTAN	0.80	ARMENIA	0.69
UKRAINE	0.73	AZERBAYDZHAN	0.80	ESTONIA	0.67
TADZHIKISTAN	0.73	KIRGIZIYA	0.78	KIRGIZIYA	0.66
MOLDAVIA	0.71	ESTONIA	0.78	TADZHIKISTAN	0.60
KIRGIZIYA	0.69	MOLDAVIA	0.74	LITHUANIA	0.60
LITHUANIA	0.66	LITHUANIA	0.70	RSFSR	0.60
UZBEKISTAN	0.62	RSFSR	0.65	AZERBAYDZHAN	0.59
RSFSR	0.62	UZBEKISTAN	0.63	UZBEKISTAN	0.57
TURKMENISTAN	0.53	TURKMENISTAN	0.53	TURKMENISTAN	0.47
KAZAKHSTAN	0.48	KAZAKHSTAN	0.51	KAZAKHSTAN	0.45
USSR	0.64		0.69		0.61

Sources: Assorted tables from GOSKOMSTAT bulletins nos. 1–7; 9–12, 1989; *Statisticheskie materialy ob ekonomicheskom i sotsial'nom razvitii soyuznykh i avtonomnykh respublik, avtonomykh oblastey i okrugov: chasti* 1 and 2; *Narodnoye Khozyaystvo SSSR* various years: *Kapital'noye Stroitel'stvo SSSR.*

show how the inter-republican economic relations are distorted by the pricing system and how the pricing system acts as a mechanism for indirect subsidies to some republics.

While the presence of subsidies to the poorer republics may qualify any conclusions concerning the 'intent' of the Soviet Government to reduce inequality across Soviet territory, it does not answer the question of whether these regions would have been better off without Soviet 'interference' in their internal economic development. A relevant question to ask, however, is how the levels of and trends in inequality in the Soviet Union compare to those in other countries and economic systems. Could large and increasing regional inequality be typical in modernizing states, and, if so, how does growing inequality in the Soviet Union compare to that experienced in other countries? Jeffrey Williamson (1965) did an international cross-section analysis in order to test the relationship between levels of economic development and regional inequality in twenty-four countries – all but one of these were market economies – and found the relationship to be that of an 'inverted U'. States progressing from early stages of economic development experienced increasing levels of inequality; but once they progressed beyond the intermediate level of development, regional inequality declined significantly.

The Soviet Union, a state in the intermediate stage of development by Williamson's criteria, should have seen its regional gaps already on the decline. This has not yet happened, which suggests that, despite the recognized subsidization of the poorer regions, the Soviet system may have delayed the equalization of regions despite any well-intentioned efforts. This conclusion, however, remains questionable for several reasons. In his study of the twenty-four countries, for instance, Williamson found that a state's size was a significant dampening variable for reducing regional inequality; states that were considerably larger than others at comparable levels of development had 'significantly more serious [problems of regional inequality] than the average of their income class'. (1965, p.15) In addition to the size factor, few states have had the tumultuous history the Soviet Union has had since its Revolution, nor is there a state as diverse culturally, with so many of its constituent nations residing in its ancestral homelands as is currently the case. This last fact inevitably creates conflicts between the central government's policy initiatives and the wishes of the republics; it clouds the causes for failed initiatives, including the attempts to reduce inequality.

With the centrally planned mechanism soon to be dismantled, the final verdict on its performance in reducing inequality remains vague. Regardless of whether or not a different system would have done better in developing the relatively backward regions of the Soviet Union, the great levels of inequality in such an impoverished state remains far more politically significant than similar gaps in wealthier industrialized states. In wealthier states, the basic needs of populations can be met. The same cannot be said for the Soviet Union and any serious attempt at real economic reform, as the following section will outline, is likely to widen the gap between rich and poor.

Perestroyka and implications

While the primary objective of the Gorbachev reform program is to bring efficiency, growth and a greater degree of rationality to the Soviet economy, many of the program's constituent parts, and the implicit changes they will require and engender, while helping some regions, are likely to have deleterious effects on others. Changes underway before the Gorbachev era were responsible for fostering noticeable spatial shifts in the economy (Dienes, 1989). These have become more pronounced and will inevitably continue to become more important as the reforms take a more definite shape. The current level of uncertainty on issues as fundamental as the very future of the state, or 'union', makes projections about specific components of the reform package not only difficult, but perhaps also academic.

The vacuum created by the partial dismantling of the highly centralized institutions of Soviet rule has not yet been replaced with functioning new institutions of government. One noticeable result of this time lag, exacerbated by the rather rapid decline in credibility of the Communist Party, has been the absence of a unifying force strong enough to counter the centrifugal forces threatening the unity of the state. The ongoing discussions over the terms of federation or the 'new union' reflect the deep-rooted animosities among the country's disparate regions and nations that make it hard to imagine the creation of an acceptable 'new' union.[30] While most analysts have explained the crisis of union largely in terms of ethnic animosities and the so-called 'Nationalities Question', many declarations of sovereignty and several separatist movements have had less to do with nationalism than with economic and political control over territory. Peoples in at least thirty-eight territorial units have issued declarations of some sort of sovereignty or independence, several of them predominantly of the same nationality as the republics from which they are seeking greater independence.

The outcome of the Union Treaty debate will have important consequences for economic relations among republics *if* the new union truly differs from the old. The extent to which the central authorities will continue to determine the financial terms between the centre and the republics (and therefore the spatial allocation of resources) will surely depend on what kind of union is created. Some views on the union treaty will provide a background to the broader question of how the reforms are likely to influence development trends across Soviet territory.

The union treaty

Clearly, a major result of the reforms thus far has been the free expression of dissent and political views across the Soviet Union. The impact of the right to express one's opinions, while perhaps being greater than President Gorbachev imagined in 1985, has forced the Soviet political agenda faster than the existing political institutions could control. As a result, long-held dissatisfaction on the part of virtually all the state's constituent republics and many sub-republican administrative units has forced the Soviet President to release his version of the new Union Treaty earlier than he had wished. The largely negative reaction to the President's draft treaty reveals the great differences in opinion over power sharing among constituencies across the country.

President Gorbachev's initial Union Treaty proposal attempts to broaden the powers of the republics at large, but the changes offered fall well short of what many of the republics want, especially in the sphere of economics. While there are differences among republics in terms of what each seeks before it claims it will 'join' the new union, the largest and wealthiest republics – the RSFSR, Ukraine and Belorussia – plus, at the least, the Baltic Republics, Georgia and Moldavia, all want greater control over their natural resources, hard currency earnings, enterprise profits, the setting of their budgets and the power to tax, than the Gorbachev proposal allows. In fact, throughout the late summer and early fall discussions on the Union Treaty, the RSFSR, Belorussia and the Ukraine wanted to limit union powers to the areas of environmental protection, defence, state security and atomic energy (FBIS-SOV-90-174, p. 31). Other republics, such as Kirgiziya, proposed that the union should also have control over pipelines, transportation and energy. President Gorbachev's proposal,[31] released in late November 1990, offered the republics seemingly greater powers; but for almost every right accorded to the republics there is an added caveat that serves to circumscribe the would-be powers of the republic. For example, while claiming in Article 8 that 'The republics are the owners of the land, its subsoil and other natural resources on their territory . . . ' the remainder of the article minimizes such ownership by adding 'with the exception of that portion of it that is essential for the implementation of the powers of the USSR'. And, 'The republics' legislative regulation of relations of ownership of land, its subsoil, and other natural resources must not impede the implementation of the Union's powers.' These powers of the USSR, described in Article 5 ('Powers of the Union'), confirm the primacy of the union over the republics in economic matters. Although section four of Article 5 uses phrases such as 'conjointly with the republics' (*sovmestno s respublikami*) to describe the move toward an all-union market, it is clear that the centre, in President Gorbachev's view, will retain the most important powers of economic decision-making, leaving the republics with little of what they claimed was necessary to keep the union together. Article 5 reads:

> Parties to the Treaty invest the USSR with the following powers: 4) the determination, conjointly with the republics, of the strategy for the country's economic development and the creation of conditions for the development of the all-Union market; the conduct of a unified financial, credit, and monetary policy based on a common currency; the compilation and execution of the union budget; the preservation and the utilization – coordinated with the republics – of gold reserves and the diamond stocks . . .

The republics' desires to control their own banking institutions and to enter freely into international agreements, which together form the basis for independent paths to development, are countered by Article Four of the treaty, under section two ('Structure of the Union'), which states: 'It is incumbent on republics . . . not to conclude agreements contrary to the aims of the Union or directed against the interests of republics belonging to it.' In short, the Gorbachev proposed union treaty is unlikely to gain confirmation by many republics without major revisions or without severe pressure from the centre. The Baltic republics have already claimed they would not even consider the president's union treaty proposal, and early reactions from most republics have not been positive.

It is unclear whether President Gorbachev's apparent shift to the right is a temporary retreat, intended to try to help bring order to Soviet society, or whether the move is yet another example of Russian and Soviet democratic reforms falling short of their goals. Despite these uncertainties, one can look beyond the potential (and specific) changes in the Soviet federation as defined by a new Union Treaty and offer some guarded forecasts about regional inequality should the broad tenets of the Gorbachev reform program, including the changeover to a market economy, continue.

Spatial implications of economic reform

Any changeover from a highly centralized allocative mechanism to a more decentralized, market-driven mechanism will undoubtedly affect regions quite differently. Under the system that dominated capital investment decisions from 1956 until the present, all-Union ministries and the all-Union GOSPLAN (State Planning Agency) maintained virtually full control over where resources were allocated. Individual republican GOSPLANs filed detailed economic plans and made specific investment requests, but actual investment decisions were made in Moscow for virtually all 'productive' and large industrial projects. Only municipal services and 'non-productive' investments, which routinely make up a small portion of the state's yearly totals, were determined within republics and smaller localities. A change in the relationship between Moscow and the republics, primarily the actual diminution of the centre's power, would be certain to affect how and where resources will be allocated. There is by no means, however, a unified position among the republics – or even among constituent regions of the larger republics – on how economic and political reform should change the current allocative and budgetary systems.

Under a more market-oriented system the poorer regions of the country, long looked upon as a drain on the wealthier regions, will undoubtedly lose whatever benefits the centralized socialist system has provided in terms of allocating capital for ideological rather than for purely economic reasons. That the southern tier received, on average, less investment capital than the western regions does not mean that the regions did not benefit from the highly centralized control of resource allocations. Had economic criteria been the sole factor in determining the allocation of resources under the centralized allocative mechanism, it is virtually certain that the LDRs would have received fewer funds.

Gorbachev's early emphasis on modernizing and expanding the existing industrial base instead of building new facilities (Dienes, 1989), if continued, should result in smaller shares of state-controlled investment for the LDRs than they have received during the past two decades. Even before the current budget deficit was 'discovered' and openly discussed, the volume of productive capital was to level off and actually decline in several regions, including labour-rich but employment-poor Central Asia and Transcaucasia. Due to the *spatial* realities of the existing Soviet industrial infrastructure, this change in investment strategy means that investments will become more concentrated in the developed western economic regions of the state at a time when the *demographic* realities of the Soviet state require greater investments to the less developed southern tier. The Soviets, however, see the western regions of the Soviet Union as 'locomotives' of development for the rest of the country (Dienes, 1989) due to their relatively advanced levels of technology and overall infrastructure, and their relatively skilled labour force (Sokolov, 1989). With a decline in investment outside these target regions, underemployment and unemployment are bound to increase in the southern tier; the number of job opportunities will not meet the demand due to the rapid population growth of the region and the extremely low incidence of out-migration among the indigenous ethnic groups.

The full implementation of new accounting procedures for all enterprises (*khozraschet*) will have regional-specific consequences, and is likely to be more harmful to the less-developed regions of the state than to the MDRs. According to this new cost-accounting system, enterprises that cannot operate profitably will face bankruptcy and possible closure. While it is uncertain how profitability will be estimated without meaningful price reform and an untangling of the vast system of subsidies for Soviet enterprises,[32] it is clear that enterprises outside the European core of the state will be the most vulnerable to any kind of performance indicators. Regional differences in capital and labour productivity make it almost certain that if subsidy support were eliminated (as *khozraschet* in effect requires), the southern tier would suffer the most. The political implications of vast closures of enterprises, especially in the light of the growing demand for jobs in the region, would be great. Even with a quiescent population, and greater economic autonomy or even independence, inequalities between northern and southern regions are sure to increase; the southern republics have narrowly developed economies and remain dependent on the rest of the Soviet Union for many of their needs.

The decentralization of political power and the concomitant shift in economic decision-making already underway in many of the union's fifteen republics is a fourth aspect of perestroyka that, if allowed to continue, is likely to widen income and developmental gaps and hurt the poorer republics.[33] Although the specifics of the republican autonomy movements remain poorly-defined, the direction and momentum are clear. Thus far, the republics' most important goals revolve around the desire to choose individual paths to development and to control a greater portion of economic production within their territories. While the *economic* irrationality of the rigid, highly centralized system of economic decision-making has finally become apparent to Gorbachev, it is important to raise the point once again that the highly centralized economic mechanism may have fostered a greater equalization of economic opportunities across Soviet territory than would have been possible without strong centralization. The change to a more decentralized system may eventually bring greater efficiency to whatever is to remain of an all-Union economy, but it will also eliminate many of the features that prevented large social and economic gaps among regions from growing even wider. If the new Union Treaty allows each republic to set its own development plans, choose to invest its earned income as it pleases, establish its own tax system, and be free to make economic agreements with foreign countries and firms as the republics have requested, income inequality and developmental gaps are likely to increase – especially in the short-term.

These measures, while creating problems for the less-developed republics, would eventually help the developed regions and make them the 'financial winners' of the economic reorganization associated with perestroyka. The Soviet leadership, and those who still favour 'union,' are caught between the demands for granting real autonomy to the republics and their titular nations, and the realization that giving up economic and political control would inevitably lead to the complete dissolution of central authority. Making a choice, however, may no longer be an option. The momentum of the independence movements, fuelled by the power of nationalism, may force the creation of a federation of fifteen truly sovereign entities, held together only by economic necessity, with the power of the centre limited to defence and foreign affairs.[34] While such a development may have been unacceptable to Mikhail Gorbachev and defenders of the system as recently as 1988 or even early 1989, recent developments may have already eliminated most options for the Soviet President without the use of brute force. Economic sovereignty or independence would possibly shift the burden of public dissatisfaction over worsening social and economic conditions from the centre to the individual republican governments, which may appear (fleetingly) appealing to those favouring the maintenance of the union. However, the kind of instability that is likely to follow significant economic and political reorganization, no matter who gets the blame, has to be minimized if the union – even in the loosest of forms – is to survive.

The failure of the Soviet system to meet its major goal of reducing inequality may have contributed to the possible disintegration of the Soviet Union. It is unclear, however, whether the new political entities that could emerge from the Soviet Union would be able to overcome their dependence on the former and remaining Soviet republics, or do better than trading one dependency for another – a new dependence on western industrialized states looking for relatively cheap labour and resources. It is clear that after seventy years of socialist rule, inequalities among regions and nations across the Soviet Union persist and have increased during the past few decades. The political implications of these trends have contributed to what is now going on in the streets and parliaments of Moscow, Vilnius, Tallinn, Kiev, Alma-Ata and many other cities across the Soviet Union. The combined effects of these movements may soon alter the map of the Soviet Union and strengthen the movement towards a new political order, begun, ironically, in the neighbouring socialist states of Eastern Europe.

Notes

1. Some regional economists, like Jonathan Schiffer, 1989, argue persuasively that equalization was never a primary objective of regional planning decision-making, since the majority of investment decisions rested with planning bodies that were dominated by all-Union ministries. These ministries, Schiffer argues, formulate their investment policies first and foremost on estimates of profit; they care little about equalizing the social and economic profiles of Soviet regions. Other Soviet specialists, however, give greater attention to the political power of the regions in explaining resource allocations. See Bahry, 1987.
2. For an early account of the Gorbachev reform program and the necessary changes in the terms of the social contract, see Hauslohner, 1987.
3. For somewhat differing views on this issue, see

Schiffer, 1989 and Bahry, 1987.

4. For a good pre-1985 summary of works on equalization, see Schiffer 1985.
5. Work by Liebowitz, 1987, showed how trends in investment allocations differed by changing the scale of analysis from the fifteen union republics to more than one hundred oblast-level units.
6. The question of reliability forces the researcher to go beyond the 'standard' problems of working with Soviet-published data. By standard problems I mean those problems that stem from (1) an artificial pricing system, which affects regions and republics differently depending on the economic structures of their economies; (2) different definitions for the same phenomena among republican statistical collection administrations; (3) territorial changes and others. See Treml and Hardt, 1972.
7. The 'conditions' or variables used in this study are a function of what data are currently available for interrepublic comparisons.
8. Unfortunately, at this juncture of my research, I must rely on republic level data, although as more data become available I plan to do much of what follows on the more detailed oblast level.
9. The unweighted coefficient of variation (CV) measures the dispersion of a particular measure relative to the group average (a group's standard deviation divided by its mean). The weighted CV does the same but each unit in the group is weighted by its share of the all-Union population.
10. The quintile ratio (QR) is the ratio of the median scores of the first and fifth quintiles. It is a good complement to the CV because, unlike the CV, it focuses more on absolute differences between top-ranked and bottom-ranked scores than on the overall dispersion among all scores in the sample. It is a good measure of range.
11. The move toward a market economy and the proposal to replace the word 'socialist' with 'sovereign' in the state's name would seem to be strong evidence that the country is moving away from socialism.
12. The most common publications to publish the press bulletin data have been *Moskovskiye Novosti*, *Ekonomika i Zhizn'*, *Izvestiya*, and the monthly *Vestnik Statistiki*.
13. The following comparisons are based on data from: Goskomstat bulletins Nos. 1–7; 9–12, 1989 and No. 16, 1990; *Statisticheskie materialy ob ekonomicheskom i sotsial'nom razvitii soyuznykh i avtonomnykh respublik, avtonomykh oblastey i okrugov: chast' 1 i chast' 2; Narodnoye Khozyaystvo SSSR v 1987g*, and *1988g; Trud v SSSR; Naseleniye SSSR 1987* and *Naseleniye SSSR 1988; Promyshlennost' SSSR; Kapital'noye Stroitel'stvo SSSR;* and *Sel'skoye Khozyaystvo SSSR.*
14. Central Asia refers to the four republics that comprise the Central Asian economic region – the Uzbek, Tadzhik, Turkmen and Kirgiz SSRs. Transcaucasia refers to Armenia, Azerbaydzhan and Georgia. Collectively, along with Kazakhstan, these republics will be referred to as the 'southern tier' and the non-Slavic south.
15. For a bibliography of works in English and Russian on the role of the second economy in the Soviet Union, see Grossman, 1990.
16. One must view these rouble values with caution since the value of what is produced is dependent on a valuation system that does not reflect world prices or even the real costs of production within the Soviet Union. For a review of the problems of analysing the valuation of natural resources as part of the debate over regional development, see Schiffer 1989.
17. Per capita production of national income for the Russian Republic in 1988 was 2769 roubles, for Uzbekistan it was 1185 roubles, and for Tadzhikistan it was 1033 roubles. Figures for 1980 were, respectively, 1969, 1083 and 1013 roubles.
18. For a good review of the regional differences in Soviet agricultural production, see Evans, 1981.
19. One need only study the movements in IMRs between 1970 and 1975 in order to see why these data need to be viewed with healthy scepticism: infant mortality was reported to have increased from 45.9/1000 to 80.8/1000 in Tadzhikistan between 1970 and 1975, and then declined to 58.1/1000 five years later, while in Uzbekistan, the IMR increased from 31.0/1000 to 53.8/1000 from 1970 to 1975. For a good review of infant mortality data and explanations for their erratic nature, see Jones and Grupp, 1983.
20. Part of this difference is a function of the vastly different IMRs, which influences life expectancy estimates.
21. For example, in 1988, DRs in Uzbekistan and Tadzhikistan were 6.8 and 7.0 per 1000 population, while those in Latvia and the Ukraine were 12.1 and 11.7 per 1000 for the same year.
22. The burden on local budgets for social services has increased over the years. See Bahry 1987. In the mid-1950s, local budgets financed approximately 75 per cent of expenditures on health; by 1980, the figure approached 90 per cent, and in 1987 and 1988 it increased to more than 94 per cent.
23. The base population for this variable was standardized so that differences in the age structures of the republics' populations would not influence the ratios.

Graduates with higher degrees were taken as a ratio of the population old enough to have completed higher education instead of as a ratio of the republic's total population.

24. These numbers represent the number of graduates per 1000 population, aged twenty and older. Using the twenty and older base population may favour those republics with younger populations: it was more difficult for the older generations to complete higher education than it is today. This is especially the case in the Baltic republics, where a significant percentage of the population did not go through the Soviet educational system and therefore few of the elderly in these republics could have completed a higher education.
25. The problems of geographical scale of analysis and the definition of 'equality' have been mentioned earlier in this paper and elsewhere. An oblast level analysis for 1956–75 has been done (Liebowitz, 1987) and a later version of this paper will include an updated oblast level analysis.
26. These questions are raised and their arguments addressed in Gillula, 1979, and Schiffer, 1989.
27. Except for the Turkmen SSR, which has received large amounts of industrial investment due to natural gas development, the southern tier has received, on average, a far smaller percentage of its total investment for industrial development. See Liebowitz 1990.
28. Reports of large-scale Russian out-migration from the southern tier go back to 1988. Recently, TASS reported that 60,000 Russians ('Russian-speaking population') will leave Kirgiziya this year, and added that since 70 per cent of the industrial workers, engineers and builders are Russophone, the outmigration of non-indigenes will have great consequences. See FBIS (SOV-90-166), 27 August 1990, p. 117.
29. See Gillula, 1979; Schiffer, 1989.
30. See Kaiser, Chapter Three of this volume, for a fuller discussion on the new union treaty and the Soviet federation.
31. For the text of Gorbachev's union treaty proposal, see the 24 November 1990 edition of *Izvestiya*, pp. 1–2. For an English translation, see FBIS-SOV-90-227, 26 November 1990, pp. 39–42.
32. Without a rational pricing system it is nearly impossible to identify which enterprises are solvent and which are not. Some resource-related industries, for example, may appear to be operating at a loss because the price structure of their products is artificially low. Similarly, some customer goods enterprises may appear to be profitable as a result of the overvalued products they supply.
33. See A. Bond, 1987.
34. Such a scenario assumes President Gorbachev and conservative forces will not resort to force in order to gain a more unitary, federal union. See Kux, 1990. For the relationship between the federal structure and nationalism, see Gleason, 1990.

Suggested reading

Bahry, D. (1987), *Outside Moscow: power, politics and budgetary policy in the Soviet Republics*, Columbia University Press, New York.

Bandera, V. and Melnyk, L. (eds) (1973), *The Soviet Economy in Regional Perspective*, Praeger, New York.

Dienes, L. (1989), *Soviet Asia: Economic Development and National Policy Choices*, Westview Press, Boulder.

Lewis, R.A. (ed.) (1990), *Geographical Studies on Central Asia*, Unwin and Hyman, London.

Schiffer, J. (1989), *Soviet Regional Economic Policy: the East-West Debate over Pacific Siberian Development*, St. Martin's Press, New York.

References

Bahry, D., Nechemias, C. (1981), 'Half full or half empty? the debate over Soviet regional inequality', *Slavic Review*, 40, pp. 366–83.

Bahry, D. (1987), *Outside Moscow: power, politics and budgetary policy in the Soviet Republics*, Columbia University Press, New York.

Bond, A. (1987), 'Spatial dimensions of Gorbachev's economic strategy', *Soviet Geography*, 28 (7), pp. 490–523.

Brada, J. (1990), 'Socialist economic integration: achievements and prospects', in *Perestroika and East-West Economic Relations: Prospects for the 1990s*, Kraus, M. and Liebowitz, R.D. (eds), New York University Press, New York.

Clayton, E. (1975), 'Regional consumption expenditures in the Soviet Union', *ACES Bulletin* 17, pp. 33–43.

Connor, W. (1984), *The Nationalist Question in Marxist-Leninist Theory and Strategy*, Princeton University Press, Princeton.

Dienes, L. (1987), *Soviet Asia: Economic Development and National Policy Choices*, Westview Press, Boulder.

Dienes, L. (1989), 'Perestroyka and the Slavic regions', *Soviet Economy* 5 (3), pp. 251–75.

Dyker, David (1983), *The Process of Investment in the USSR,* Cambridge University Press, Cambridge.

Evans, A. (1981), 'Interrepublic inequality in agricultural development in the USSR', *Slavic Review,* 40, pp. 570–84.

Gardner, S. (1990), 'The implications for greater East-West economic cooperation for the Soviet economy', in *Perestroika and East-West Economic Relations: Prospects for the 1990s,* Kraus, M. and Liebowitz, R.D. (eds), New York University Press, New York.

Gillula, J. (1978), 'Regional interdependence and economic development in the USSR: an interregional input-output analysis', unpublished Ph.D. dissertation, Duke University.

Gillula, J. (1979), 'The economic interdependence of Soviet Republics, in US Congress', *JEC,* USGPO, Washington, D.C., pp. 618–55.

Gleason, G. (1990), *Federalism and Nationalism: the struggle for Republican Rights in the USSR,* Westview Press, New York.

Goskomstat SSSR (1989), (1990), *Satatisticheskiy Press Byulleten'* Nos. 1–7, 9–12 (1989) and No. 16 (1990), Informatsionno-Izdatel'skiy tsentr, Moscow.

Grossman, G. (1990), *The Second Economy in the USSR and Eastern Europe: a bibliography,* Paper No. 21, the Berkeley-Duke project on the second economy in the USSR.

Hauslohner, P. (1987), 'Gorbachev's social contract', *Soviet Economy,* 3 (1), pp. 54–89.

Holubnychy, V. (1973), 'Spatial efficiency in the Soviet economy', in *The Soviet Economy in Regional Perspective,* Bandera, V. and Melnyk, L. (eds), Praeger, New York.

Huzinec, George (1977), 'A reexamination of Soviet industrial location theory', *The Professional Geographer,* 29 (4), pp. 259–65.

Jones, E. and Grupp, F. (1982), 'Modernization and equalization in the USSR', *Soviet Studies,* 36, pp. 159–84.

Jones, E. and Grupp, F. (1983), 'Infant mortality trends in the Soviet Union', *Population and Development Review,* 9 (2), pp. 213–46.

Koropeckyj, I.S. (1972), 'Equalization of regional development in socialist countries: an empirical study', *Economic Development and Cultural Change,* 21, pp. 68–86.

Koropeckyj, I.S. (ed.) (1977), *The Ukraine within the USSR,* Praeger Press, New York.

Kux, S. (1990), 'The Ukraine within the USSR', *Problems of Communism,* XXXIX, March–April, pp. 1–20.

Liebowitz, R. (1987) 'Soviet investment strategy: a further test of the "equalization hypothesis"', *Annals of the Association of American Geographers,* 77 (3), pp. 396–407.

Liebowitz, R. (1990), 'Soviet geographical imbalances and Soviet Central Asia', in *Geographical Studies on Soviet Central Asia,* Lewis, R.A. (ed.), Unwin and Hyman, London.

McAuley, A. (1979), *Economic Welfare in the Soviet Union,* Allen and Unwin, London.

Melnyk, L. (1977), 'Capital formation and financial relations', in *The Ukraine within the USSR,* Koropeckyj, I.S. (ed.), Praeger Press, New York, pp. 268–99.

Mezhevich, M. (1978), 'Kompleksnoye planirovaniye krupnykh gorodov', *Planovoye Khozyaystvo* 3, pp. 110–15.

Mil'ner, G. and Gilinskaya, E. (1975), 'Interregional regulation of the population's living standard', *Problems of Economics,* XVIII, December, pp. 55–63.

Mil'ner, G.V. (1969), 'Urovni' zhizni naseleniya Dal'nego Vostoka i osnovnye napravleniya ego povysheniya', in *Vosproizvodstvo trudovykh resursov Daln'ego Vostoka,* L. Rybakovskii (ed.), Nauka, Moscow, pp. 104–16.

Mil'ner, G.V. (1979), 'Problems of ensuring the supply of labor resources for Siberian and Far Eastern regions', *Problems of Economics,* XXII, June, pp. 82–97.

Mints, A.A. (1976), 'A predictive hypothesis of economic development in the European part of the USSR', *Soviet Geography,* 17, pp. 1–27.

Morton, H. (1983), 'Local services and the attempt to rationalize the delivery of urban services, in *Soviet Local Politics and Government,* Jacobs, E. (ed.), Allen and Unwin, London, pp. 186–203.

Nahaylo, B. and Swoboda, V. (1990), *The Soviet Disunion: a History of the Nationalities Problem in the USSR,* Free Press, New York.

Nechemias, C. (1980), 'Regional differentiation of living standards in the RSFSR: the issue of inequality', *Soviet Studies* 32, pp. 366–78.

Nove, Alec (1978), *The Soviet Economic System,* Allen and Unwin, London.

Rumer, B. (1989), *Soviet Central Asia: a Tragic Experiment,* Unwin and Hyman, Cambridge.

Rutgayzer, V. (1974), 'National economic assessment of the activity of branches in the non-productive sphere', *Problems of Economics,* XVII, October, pp. 44–61.

Rybakovskiy, L.L. (ed.) (1969), *Vosproizvodstvo trudovykh resursov Dal'nego Vostoka,* Nauka, Moscow.

Schiffer, J. (1985), 'Interpretations of the issue of "inequality" in Soviet regional policy debates', *International Journal of Urban and Regional Research,* 9 (4), pp. 508–22.

Schiffer, J. (1989), *Soviet regional economic policy: the*

East-West debate over Pacific Siberian development, St. Martin's Press, New York.

Schroeder, G. (1973), 'Regional differences in incomes and levels of living in the USSR', in *The Soviet Economy in Regional Perspective,* Bandera, V. and Melnyk, L. (eds), Praeger, New York.

Schroeder, G. (1981), 'Regional living standard', in *Economics of Soviet Regions,* Koropeckyj, I. and Schroeder, G. (eds), Praeger, New York.

Selyunin, V. and Khanin, G. (1987), 'Lukavaya Tsifra', *Novyy Mir,* (2), pp. 181–201.

Serdit, N. (1969), 'Trudovye resursy, na zanatye v obshchestvennom khozyaystve', in *Vosproizvodstvo trudovykh resursov Dal'nego Vostoka,* Rybakovskiy, L.L. (ed.), Nauka, Moscow.

Sokolov, V. (1989), 'Kontseptsiya', *Literaturnaya Gazeta,* 31, 2 August.

Spechler, M. (1979), 'Regional developments in the USSR, 1958–1978', in US Congress, *JEC,* USGPO, Washington, D.C., pp. 141–63.

Topilin, A.V. (1975), *Territorial'noye pereraspredeleniye trudovykh resursov v SSSR,* Ekonomika, Moscow.

Treml, V. and Hardt, J. (eds), (1972), *Soviet Economic Statistics,* Duke University Press, Durham, N.C.

USSRa (various years), *Gosudarstvennyy Byudzhet SSSR,* Finansy i Statistika, Moscow.

USSRb (1972), *Itogi Vsesoyuznoy Perepisi Naseleniya 1970 goda, III tom,* Statistika, Moscow.

USSRc, 1988, *Kapital'noye Stroitel'stvo SSR,* Finansy i Statistika, Moscow.

USSRd (various years), *Narodnoye Khozyaystvo SSSR,* Finansy i Statistika, Moscow.

USSRe (1988), (1989), *Naseleniye SSR,* Finansy i Statistika, Moscow.

USSRf (1988), *Promyshlennost' SSR,* Finansy i Statistika, Moscow.

USSRg (1988), *Sel'skoye Khozyaystvo SSSR,* Finansy i Statistika, Moscow.

USSRh (1988), *Trud v SSSR,* Finansy i Statistika, Moscow.

Vestnik Statistiki (1990), 'Obem vvoza i vyvoza produktsii po soyuznym respublikam za 1988g vo vnutrennykh i mirovykh tsenakh', 4, pp. 49–60.

Wagener, H.J. (1973), 'Rules of location and their concept of rationality', in *The Soviet Economy in Regional Perspective,* Bandera, V. and Melnyk, L. (eds), Praeger, New York.

Williamson, J. (1965), 'Regional inequality and the process of national development', *Economic Development and Cultural Change,* XIII (4), pp. 3–84.

Woroniak, A. (1973), 'Regional aspects of Soviet planning and industrial organization', in *The Soviet Economy in Regional Perspective,* Bandera, V. and Melnyk, L. (eds), Praeger, New York.

Zakumbayev, A.K. (1977), *Ekonomicheskoye razvitiye soyuznykh respublik i rayonov,* Nauka, Alma-Ata.

3

Nationalism: the challenge to Soviet federalism

Robert J. Kaiser[1]

Introduction

The contemporary nationalist challenge to the Soviet federation is one of the most significant problems to confront the political leadership during the post-war period. Since 1985, in large part as an unintended consequence of the reforms implemented under Gorbachev, the centre's ability to obtain compliance through coercive means or even the threat of coercion has been severely restricted. In this new environment, for the first time since the formative years of the Soviet Union, nationalism has become overtly political. It remains to be seen whether or not a new Soviet federation or confederation can be created that will accommodate the nationalist demands for greater self-determination while retaining some *raison d'etre* for the Union, or alternatively whether the central authorities will attempt to reinstitute coercive mechanisms to restore law and order.

An intimate relationship exists between national communities and the federal structure of the state. The federal units (union republics, autonomous republics, autonomous oblasts, autonomous okrugs) were delimited, at least in part, on the basis of national homelands, and have become the territorial bases for nationalist claims to sovereignty today. From a system instituted from above to help manage the national problem, Soviet federalism has become an instrument used from below to promote indigenous nationalist ambitions for self-determination. This pressure for greater 'indigenization' from below has had a number of serious consequences for the Soviet Union, and forms the central theme of this chapter.

This chapter focuses upon nationalism and federalism in the fifteen union republics, since it is this tier in the federal hierarchy which is said to be sovereign and which has the constitutional right of self-determination, up to and including secession. The lower tiers – the autonomous republics (ASSRs), autonomous oblasts and autonomous okrugs – are considered only peripherally (Map 3.1 provides the federal administrative structure of the Soviet Union).

The evolving role of federalism in solving the national problem

Soviet federalism, as it existed before to 1985, was in many ways more apparent than real. The political administrative structure of the Soviet Union has been described by western analysts as federal in form but unitary in essence (Pipes, 1968; Connor, 1984; Gleason, 1990), and by the Soviet Constitution as an 'integral, federal, multinational state' (Shevtsov, 1982, p. 7). In the West, the so-called 'totalitarian model' has been used as a contextual framework within which to assess Soviet interstate and domestic political behaviour from the Stalin era to the 1980s, indicating that the federal structure has been viewed as a hollow shell in a political system where power is concentrated at the top.[2]

At the other end of the political administrative structural continuum lies the confederal state, with each union republic an essentially sovereign nation-state. Certain features of the 1922 Union Treaty and the subsequent Soviet Constitutions, such as the right of union republics to enter into foreign relations, to coin money, the right to their own constitution, flag and anthem, and perhaps most importantly their right to self-determination, up to and including secession, have given the Soviet Union the appearance of a confederation. However, the concentration of political power in a hierarchically structured and highly centralized Communist Party, in addition to the use or threat of coercion should nationalists attempt to exercise their 'sovereign' rights, have in the past offset the confederal aspects of the Soviet Constitution. While not arguing in favour of the totalitarian model of Soviet society, it must nonetheless be acknowledged that Soviet federalism has until recently represented a much more centralized, unitary political system than one normally associates with a federated state.

The limited essence of Soviet federalism is related to Marxist-Leninist attitudes toward this type of state, as well as the evolution of the Communist Party's approach to solving the so-called 'national problem'. Before the October Revolution and the chaos that followed, Lenin was in theory opposed to federalism, particularly in multinational states such as Russia; he favoured instead large, unitary states (Lenin, 1913, p. 38):

> But while, and insofar as, different nations constitute a single state, Marxists will never, under any circumstances, advocate either the federal principle or decentralization. The great centralized state is a

Map 3.1 *The territorial administrative system of the USSR*

Key to Map 3.1

1 Armenian
2 Azerbaydzhan
3 Belorussian
4 Estonian
5 Georgian
6 Kazakh
7 Kirgiz
8 Latvian
9 Lithuanian
10 Moldavian
11 Russian SFSR
12 Tadzhik
13 Turkmen
14 Ukrainian
15 Uzbek
16 Abkhaz
17 Adzhar
18 Bashkir
19 Buryat
20 Chechen-Ingush
21 Chuvash
22 Dagestan
23 Khabardin-Balkar
24 Kalmyk
25 Karakalpak
26 Karelian
27 Komi
28 Mari
29 Mordvin
30 Nakhichevan
31 Severo-Ossetin
32 Tatar
33 Tuvin
34 Udmurt
35 Yakut
36 Adyge
37 Gorno-Altay
38 Gorno-Badakhshan
39 Karachay-Cherkess
40 Kakhass
41 Nargorno-Karabakh
42 Evrei
43 Yugo-Ossetin
44 Aga Buryat
45 Chukchi
46 Evenki
47 Khanty-Mansi
48 Komi-Permyak
49 Koryak
50 Nenets
51 Taymyr
52 Ust Orda
53 Yamalo-Nenets

tremendous historical step forward from medieval disunity to the future socialist unity of the whole world, and only via such a state . . . can there be any road to socialism.

After the October Revolution and the failure of a spontaneous world socialist revolution to occur during the First World War, this attitude changed. While still opposed to federalization of the state in principle, a federal system was seen as a necessary expedient to overcome the national problem that the Bolsheviks had inherited from the Russian Empire. According to Lenin (1914, p. 199), the national problem in large part resulted from the unequal treatment of the non-Russian nations in the state. The system of national stratification in the late nineteenth century Russian Empire had placed Russians in a dominant position, and Russification of the non-Russian population in the state, or alternatively their treatment as '*inorodtsy*' (foreigners) was actively promoted by the Tsarist autocracy.[3] To overcome the legacy of this 'Great Russian chauvinism', international equalization would be required, and a federal system which would guarantee political equality between the major nations in the state was justified in part on this basis (Lenin, 1922, pp. 719–24).[4] With this purpose in mind, the federal units were delimited at least in part according to the geographic extent of ethno-national communities, which in turn were rough approximations of the historic homelands of these groups. Federalization primarily on the basis of ethno-territorial communities gave the appearance of nationalizing territory, and this was enhanced by the confederal nature of the 1922 Union Treaty and the Soviet Constitutions.

Thus, the federal system became acceptable to the Communist Party leadership at least in part because it was seen as a necessity in solving the national problem; it was therefore viewed as transitory. Once international inequalities from the past had been overcome, and once the nations within the state had become developed and socialized into one Soviet people, the need for a federated state would disappear. This meant that not only was the federated state transitory, but that nations themselves were to disappear in favour of a new communal form of identity uniting the entire population of the Soviet Union. As communism succeeded in replacing capitalist systems, this new community would ultimately extend to encompass the entire human race. A communist world in theory would be a world without nations (or states or classes for that matter) – and one without international conflict.

With no socialist world revolution in sight, the Soviet Union under Stalin embarked on a path toward 'socialism in one country'. As this change in focus occurred, a major shift in nationality policy also emerged. The Communist Party leadership moved away from support for the equalization of each national community in its respective home republic (the policy of *korenizatsiya* or 'indigenization') and toward the construction of a unified Soviet people. This policy shift was not anational, but favoured the Russification of the population as a centralizing alternative to divisive indigenization. Thus, while the 1936 Constitution was written as a federal document with confederal elements, the reality of the times was that nations in the state were increasingly losing their independence of action. The use of coercion against any display of 'bourgeois nationalism' reached its zenith in the Great Purges of the late 1930s (Conquest, 1967, pp. 92–7).

During the post-Stalin period, the utility of Soviet federalism has continued to be measured in terms of resolving the national problem; as such, the system has retained a transitory appearance. For Khrushchev, the attainment of communism in the Soviet Union, slated for 1980, was to be preceded by the disappearance of nations in the country and their merger into one Soviet People (Connor, 1984, pp. 398–9). At the same time, this meant that Soviet federalism had served its transitional function of overcoming the national problem associated with inequality, and that the federal system itself had lost its importance even by the early 1960s. According to the 1961 Party Programme (Kommunist, 1961, p. 84):

> The borders between the union republics within the USSR are increasingly losing their former significance, since all nations are equal, their life is organized on a single socialist foundation, the material and spiritual needs of each people are satisfied to the same extent, and they are all united into one family by common vital interests and are advancing together toward a single goal – communism.

Events during Krushchev's remaining tenure in office contradicted this utopian image of a single communist family. When the horizontal (i.e. territorial) *sovnarkhozy* economic administrative system was established to reduce the power of the vertical ministerial structure, the decentralization of economic decision-making that accompanied this change swiftly led to a rise in national separatism in the economic realm as members of indigenous nations from around the country acted to increase their level of autarky. This indicated both that the national problem was far from being resolved, and also that the republican boundaries were far from becoming obsolete, at least from the vantage point of the nationally self-conscious indigens in the state. As a result, the experiment with economic decentralization was reversed during the last years of Khrushchev's leadership, and the return to a more vertical, ministerial structure was continued under Brezhnev, though the republics retained greater control over economic decision-making than during the Stalin era (Bahry, 1987, p. 1).

During the 1960s, a debate centring around the question of whether or not the federation had outlived its usefulness was waged in the pages of *Voprosy Istorii* (Hodnett, 1967). During the Brezhnev era there was a relative retreat from the position that the nations of the state were well on the way toward merger into one Soviet people, but the necessity of retaining the federal system continued to be debated in anticipation of the 1977 Constitution. For Brezhnev, the claim that the national question as it existed in the past had been resolved appeared to mean that international equalization had been achieved (Schroeder, 1990, p. 43). This seemed to indicate that Brezhnev also felt that the federal system had served its purpose and was no longer needed. However, the 1977 Constitution retained the federal structure of the state and the confederal rights of the union republics, and was a victory for those who argued that Soviet federalism retained its relevance even during the stage of 'advanced socialism' that the Soviet Union was said to have entered. In part, the argument in favour of continued federalism was successful because it was decoupled from the goal of international equalization, which in theory had already been achieved, and reconstructed around the goal of attaining communism (Gleason, 1990, pp. 72–3).

Nonetheless, the idea that a Soviet People has emerged, that while individuals still retain their sense of national identity they have been successfully socialized or 'Sovietized' during the course of the past seventy years, continues to make itself felt in the statements and behaviour of the Soviet leaders. Gorbachev clearly believed this to be the case; as a result, he ignored or played down the potentially disruptive consequences that perestroyka would have in the management of the national question. It was not until 1988 that Gorbachev recognized the national problem as 'one of the most fundamental, vital questions facing (Soviet) society' (Gorbachev, 1988, p. 2). By that time, Moscow had already lost the initiative in managing international relations within the state, and found itself consigned to an essentially reactive position from which it is still attempting to escape.[5]

The reform program implemented under Gorbachev has thus been founded on the faulty assumption of successful political socialization. This has resulted in a number of unforeseen or unintended consequences. The most significant of these is that nationalism rather than 'proletarian internationalism' has absorbed much of the creative potential that was unleashed by *glasnost* and *demokratizatsiya.* Rather than a successful restructuring of Soviet society, the reforms implemented in the past few years have increasingly encouraged political separatism along national lines. The reorientation of the national front organizations formed during the latter half of the 1980s is symbolic of this change. Many, such as Sajudis (the Lithuanian word for perestroyka) and RUKH (the popular movement of Ukraine for restructuring) began as organizations in support of perestroyka, but have since become political organizations for the 'rebirth' or 'reawakening' of their respective nations. As this has occurred, the political administrative system has been pushed toward the confederal end of the unitary-confederal continuum at an accelerating rate.

In searching for explanations of the contemporary and long-term trends in nationalism and federalism in the Soviet Union, it is necessary to go beyond the Kremlin walls and the centrally constructed nationality policies, and begin to explore the relationship between nationalism and Soviet federalism from the perspective of the nations in the state. A geographic approach to this topic is particularly relevant. Territory is an often-overlooked though crucial dimension of the national problem in multinational, multi-homeland states generally, which comprise a majority of countries in the world today (Bruk, 1986). In the Soviet Union, the relationship between nation and homeland is especially important, since the national homeland provided a basic geographic parameter in the delimitation of union republics, autonomous republics, autonomous oblasts and okrugs. In turn the perceptual borders of the national homeland have been shaped to some degree by the federalization of the state. The latter is particularly true for those ethnic communities that underwent a process of national consolidation relatively late. Examples of this include the Tadzhiks (Rakowska-Harmstone, 1970, pp. 76–9) and the Belorussians (Vakar, 1956; Guthier, 1977). While these international distinctions continue to make themselves felt, it is nevertheless true that nationalist claims to a homeland on which national self-determination may be exercised are a nearly ubiquitous feature of the national problem facing the Soviet Union.

The importance of homeland

An ancestral or even 'primordial' homeland is often proclaimed as a fundamental feature of the national community. The meaning of homeland to nationally self-conscious individuals is as much subjective as objective. It has to do with a deep emotional attachment to a region as the geographic cradle of the nation, as the place where the national membership experienced a glorious past, and as the place where the nation's destiny is to be fulfilled in the future (Smith, 1981 and 1986; Williams and Smith, 1983; Connor, 1986). The formation of this more subjective 'sense of homeland' is intimately related to the development of a national self-consciousness, which itself revolves around a belief in a common past and a common future together (Emerson 1960, p. 95). The way in which territory has impinged on the formation of national self-consciousness and the evolution of international relations is briefly examined in this section.

A belief in a common past refers not only to a myth about common ancestry and kinship ties uniting group members, but also to common geographic origins, i.e. a national birthplace. Indeed, the backward-looking dimension of national self-consciousness has often been presented by nationalists as a merger of 'blood and soil', and this fusion of genealogy and geography is found not only in the nationalist literature of the Romantic era over a century ago, but also in the post Second World War Soviet Union. For example, in his poem *Recognizing the Homeland*, the Uzbek Cholpan Ergash wrote (as quoted in Allworth, 1973, p. 15):

> So that my generation would comprehend the Homeland's worth,
> Men were always transformed to dust, it seems.
> The Homeland is the remains of our forefathers.
> Who turned into dust for this precious soil.

Similarly, Turar Kodzhomberdiev, a Kirgiz poet found the homeland even 'more primal than the maternal breast: " . . . Remember, even before your mother's milk / You drank the milk of the homeland."' (Allworth 1973, p. 16) This reverential attitude toward the homeland is a nearly ubiquitous phenomenon that extends to the physical features of the national landscape (Williams and Smith, 1983, p. 509):

> Its mountains are sacred, its rivers are full of memories, its lakes recall distant oaths and battles, all of which have been commemorated in national epics and ballads, and attracted countless legends. History has nationalized a strip of land, and endowed its most ordinary features with mythical content and hallowed sentiments.

Seen in this way, an indigenous nation's claim to territory must be understood as more than an objective choice of economically rational actors. Relatively worthless land in economic terms may be the scene of violent international conflicts where members of two nations see this same land as the priceless, sacred home soil. Nagorno Karabakh appears to be a contemporary case of this. It is also little wonder that protests over environmental degradation have risen along with nationalism in nearly every union republic and a number of autonomous regions of the Soviet Union in recent years, for polluting the national homeland is perceived as akin to polluting the national soul. For example, the Russian 'patriot' Yuri Bondarev recently warned (as quoted in Petro, 1987, p. 252):

If we do not stop the destruction of architectural monuments, the rape of our lands and rivers; if a moral explosion does not occur in science [and] in literary criticism, then one fine morning, which will be our last funereal [awakening], we will, with our boundless optimism, awake and understand that we have erased and forever destroyed the national culture of great Russia, her spirit, her love for the land of our fathers, her beauty, her great literature, painting, philosophy.

Delimiting the national homeland

Since the national attachment to homeland is based in part on a largely mythical belief that the nation was born in and needs a special place to thrive, the boundaries of one nation's perceived homeland are unlikely to accord well with the perceptual borders of neighbouring national homelands. Nationalists have often looked to a period in the past where their national community controlled the broadest geographic realm, which is also frequently proclaimed as the nation's historic 'golden age'. Pearson (1983, p. 17), in his study of Eastern Europe sums this up best:

What does the 'historic' argument mean? The *longest* chronological span of ownership? The *earliest* significant period of ownership? The latest or the most *beneficial* period of ownership? In practice, claimants select the criteria favouring their own case, transforming past history into present politics in the process.

As the several international territorial conflicts emerging in the Soviet Union during the past few years indicate, the republics as delimited do not precisely match the homelands as perceived by nationalists themselves. In March 1990, the US Department of State published a map cataloguing the ethno-territorial disputes that had arisen in the Soviet Union up to that time. Significantly, this map (State, 1990) is already out-of-date, since it does not include information on cross-border claims made by Estonians against the RSFSR and Latvians against the RSFSR and Belorussia; the Rumanian claim to Moldavia, and the declarations of sovereignty made by the Gagauz and Russian-speaking population in Moldavia; the rash of declarations of independence or sovereignty made throughout the Soviet Union at all levels of the federal hierarchy; the removal of Meskhetian Turks from Soviet Central Asia; and the flood of Crimean Tatars to their ancestral homeland in the Crimea. Certainly, by the time this book is published many additional new features will need to be added to the map.

Several reasons may be suggested for the lack of coincidence between perceived homeland and federal unit, including a lack of knowledge about the ethnic composition of the population in a number of outlying regions at the time of border delimitation, the lack of national self-consciousness in a number of these regions at that time, and perhaps the purposeful misallocation of regions with the intent to divide and conquer (Schwartz, 1986; Connor, 1984, pp. 302–22). There has also been a tendency for each national community to lay claim to the largest territory possible, resulting in overlapping claims to the same region. The conflictual and nearly unsolvable nature of disputes over homeland border delimitations was recently noted by a reporter visiting Moldavia (FBIS, 16 November 1990, p. 80):

There is no peace on the banks of the Dnestr . . . There are bitter calculations of who has lived where and for how many millennia. Historical documents are being dusted off, maps of population movements have been coloured, assessments of historical events revised. Dacians, goths, ancient Romans, Stephen the Great, the campaigns of Suvorov, the Russian-Turkish war, the gift of Catherine II, ancient homes, ancient townships, and so on and so forth – everything is thrown into the furnace to maintain the flames of national exclusiveness and to prove that this people alone – and no other! – have the right to this land.

As is evident throughout the Soviet Union, the use of the 'historic argument' in the contemporary struggle for territorial control is a nearly ubiquitous feature of nationalism today. In tracing the nation's past, claims to a larger homeland have been made by a number of groups. One example of this is the claim made by some Lithuanian nationalists that Kaliningrad Oblast, formerly part of East Prussia (Konigsberg) is a part of the Lithuanian homeland. The claim to what is being called 'Little Lithuania' is based on the belief that kinship ties between East Prussians and Lithuanians can be traced to medieval times (Gorbatov, 1990, p. 2), though the Lithuanian interest in Kaliningrad may be more geopolitical than ethno-territorial. Estonia and Latvia have both made claims to adjacent territory in the RSFSR and

Belorussia, though these claims are based not on ancient kinship ties but rather on the borders of the independent inter-war republics, which were changed after the 1940 reincorporation of the Baltic region. In the most well-known contemporary case, Armenians claim Nagorno Karabakh not simply because Armenians comprise a demographic majority in this autonomous oblast, but also because it is claimed as an intimate part of their ancestral homeland. There has been a proliferation of border and territorial disputes regarding this issue in recent years.

The claim to at least the entire union republic as the ancestral homeland has been used to deny the right of minorities living in the republics to 'secede' and establish their own union republics. This is clearly the case in Azerbaydzhan (Armenians in Nagorno Karabakh), Georgia (Abkhazians and Ossetians), Moldavia (Gagauz and Russian-speaking Slavs) and Lithuania (Poles). For example, in rejecting the declarations of sovereignty by the Gagauz and Russian-speaking population in Moldavia, President Mircha Snegur justified Moldavia's position by invoking the primordial nation-homeland myth: 'Not a single inch of the land of our ancestors to satisfy the autonomy ambitions of the newcomers!' (FBIS, 28 September 1990, p. 95). From the perspective of Moldavian nationalists, the Gagauz 'newcomers' have been in the area for only a few centuries, while the Moldavians have been there 'from time immemorial'.

An additional problem facing the Soviet Union is that the Russian sense of homeland expanded with the conquest of new ethnic territory over time. Currently, the geographic extent of the Russian homeland as perceived by Russian nationalists is also the most variable of any group in the state. While for some Russians *rodina* or *otechestvo* refers to the northwestern core region within which the Russian national community emerged, for most 'Russia' is at least equivalent to the RSFSR, and for some to the entire Soviet Union (Kristof, 1967; Szporluk, 1989 and 1990). For example, according to Solzhenitsyn (1990), a future Russia would include the contemporary RSFSR, Ukraine, Belorussia and some border regions populated primarily by Russians, such as Northern Kazakhstan. This Russian nationalist vision of a future homeland runs counter to the image of homeland held by Ukrainian, Belorussian and Kazakh nationalists, as well as nationalists from the dozens of national communities whose homelands are currently incorporated in the RSFSR. This was graphically demonstrated in mass protests in Kazakhstan and Ukraine after the publication of Solzhenitsyn's article.

The Russian nationalist claim to northern Kazakhstan, as well as to other regions where they currently comprise a majority, rely on a 'demographic argument' (Pearson, 1983). The same is true in the case of the Georgian claim to Abkhazia and South Ossetia. Since the borders of the federal units in the Soviet Union were delimited on the basis of ethno-demographic factors rather than historic claims to an ancestral homeland, it is not surprising that there are currently international conflicts over the claim to territory.

As a broad generalization, it appears that at minimum indigenous nationalists proclaim the right to control 'their own' territory, i.e. the union republics or autonomous units named after the nation. However, this territory is unlikely completely to encompass the ancestral homeland as perceived by the nationalists themselves. The question of an indigenous nation's 'sense of homeland' is thus more complicated than even the Soviet federal structure implies.

National Territoriality

The backward-looking 'sense of homeland' between nationally self-conscious individuals and their homelands may be thought of as comparable to Soja's (1971, p. 34) 'sense of spatial identity', which he identified as the first ingredient of 'human group territoriality'. The homeland is also of critical importance in the more future-oriented belief in a common destiny. This forward-looking dimension of national self-consciousness often involves an activated national territoriality, defined as an attempt by nationalists to gain control over the destiny of the nation by attaining sovereignty within the ancestral homeland. This more strategic or instrumentalist view of territoriality is in line with Sack (1986, p. 19), who has defined human territoriality as 'the attempt by an individual or group to affect, influence, or control people, phenomena, and relationships, by delimiting and asserting control over a geographic area'.

The belief in a national destiny is often said to involve the reconstruction of the nation's glorious past in the future, resulting in the establishment of a new national golden age (Smith, 1986). According to a common nationalist theme, this future Utopia will emerge once the nation has 'reawakened' and assumed its rightful dominant position in its ancestral homeland. This is implicit in the definition of national territoriality provided above, and also in Soja's second major ingredient of human group territoriality (1971, p. 34), a 'sense of exclusiveness'. This more active, future-oriented national territoriality is a nearly ubiquitous feature in the world political system (Connor, 1986), and is present in the contemporary Soviet Union not only in nationalist literature, but also in the programs of the national front organizations and the rhetoric of the nationalist leaders. The declared objective to become 'masters of their own land', 'which has been theirs from time immemorial' (Rukh, 1989) is a recurrent theme of nationalist movements that have emerged since 1985. The attempt to secure the destiny of the nation by gaining control of the home republic is also the primary justification used in the declarations of sovereignty made by the union and autonomous units during 1990. For example, the Turkmenistan declaration is justified as 'expressing the wishes of the people of Turkmenistan', and 'recognizing its responsibility for the fate of the Turkmen nation' (FBIS, 12 September 1990a, p. 107). This is clearly not an anational document which treats all citizens of the republic equally, but rather a statement that reflects the will of indigens to control their own lives in their own homeland. This same wording is found in every union republic-level declaration of sovereignty, with the exception of the RSFSR, which as a federated republic does proclaim that the 'multinational people are the repository of sovereignty and the source of state power' (FBIS, 14 June 1990, p. 102).[6]

The Soviet federal system, delimited at least indirectly on the basis of national homelands, has served over time to institutionalize and strengthen the sense that the home republic rightfully belongs to the national membership and to no one else. From the names of the republics, to the symbols of sovereignty (anthem, flag, constitution) to special rights and privileges for indigens in their respective home republics, the federal structure has reinforced both national self-consciousness and the 'sense of homeland' that exists as a fundamental feature of that self-consciousness. It could be argued that the creation of the federal system, which was meant to help resolve the national problem inherited from the Russian Empire, has instead helped ensure against its resolution. The very existence of a federation based even loosely on ancestral homelands almost guaranteed that the political context (federal in form, unitary in essence) as well as the socio-cultural context (national in form, socialist in content) of Soviet society would be transposed during the course of time. Rather than a brotherly society of internationalists, the Soviet Union has become a society of particularistic nations seeking to gain greater control over their own lives in the territory perceived to be their ancestral homelands. Such a strategy of national territoriality has been evident for at least the entire post-war period, and has resulted in the indigenization of life, at least in the union republics. The term 'indigenization' is the equivalent of the Russian *korenizatsiya*, which during the 1920s referred to a nationality policy of affirmative action targeting the indigenous nationalities for special treatment. However, the contemporary process of 'indigenization' is no longer solely a nationality policy directed from above to give indigens limited cultural autonomy, but has become a popular or mass indigenous national movement from below, with indigens seeking no less than national self-determination in all spheres of life.

Indigenization in the Soviet Union, as in other multinational, multi-homeland states, can be thought of as a process of national *separatism* through which each national community attempts to control its own life within its respective homeland. Indigenization from below may be seen as a counter to the state-sponsored programs of Sovietization and Russification of the non-Russian periphery. It is not only, or even primarily, a secessionist movement. National separatism has manifested itself broadly in the cultural, social, economic and political spheres, and has also helped to ensure that demographically each republic would remain relatively nationally homogeneous. The dimensions of this indigenization of life, and the spatial variations in their manifestation, are examined below.[7]

Inter-homeland migration

Historically, the major inter-homeland migratory trend in both the Russian Empire and the Soviet Union has been for Russians, and to a lesser extent other Slavs, to leave their core areas in search of new lands, better economic opportunities in emerging cities, and so on. For much of its history, the Soviet Union has encouraged the inter-homeland migration of peoples not only as a means of geographically balancing the supply and demand for labour, but also as a way of 'internationalising' the population and providing the environment within which a new Soviet people could be created (Bromley, 1983, pp. 355–8).[8] During the past thirty years alone, the number of people living outside their respective home republics increased from 27.6 million in 1959 to 43.4 million in 1989 (TsSU SSSR, 1962; Goskomstat SSSR, 1989). Nonetheless, most of the major nations in the state have remained highly concentrated in their respective home republics, and continue to comprise a majority of the people resident there (see Tables 3.1 and 3.2).

There are great differences in the inter-homeland redistribution of nations across time and from region to region. While the nations of the non-Slavic south (Transcaucasia and Central Asia) have become more demographically dominant in their respective home republics during the last thirty years (1959–89), the Slavic and Baltic nations have become a smaller proportion of their respective republics over time. This demographic dilution has been particularly rapid in Estonia and Latvia, and primarily results from a high level of Russian in-migration that has exceeded the rate of natural increase of Estonians and Latvians. In Lithuania, a trend toward growing demographic dominance by the indigenous nation during the period 1959–70 was reversed between 1970 and 1989 also due to increasing net in-migration of Russians. However, far from serving as a positive force in the Sovietization of the Baltic nations, this in-migration and indigenous demographic dilution have served as catalysts activating national territoriality in the region.

During the past decade, there has also been a return migration stream to the home republic for all three Baltic nations, and this has resulted in the increasing concentration of the members of these nations in their respective home republics. In addition to these returnees, the number of Estonians, Latvians and Lithuanians living outside their home republics is also declining due to natural decrease (i.e. a greater number of deaths than births), which is common to groups with an older age structure (*Argumenty i Fakty*, 1990, 33, p. 1). It is also true that assimilation of Estonians, Latvians and Lithuanians occurs primarily among those living outside their respective homelands. This is true for most other nations in the state as well (Kaiser, 1988).

The Slavic nations declined as a per cent of their respective home republics during the entire period, and this was due primarily to Slavic out-migration to other republics, plus Russian in-migration to Ukraine and Belorussia. A major reversal occurred during the 1980s, when Russian out-migration from the non-Slavic south to the RSFSR resulted in a stable percentage of Russians in their home republic for the first time since the Second World War (see Table 3.2). Since the census in January 1989, the rate of Slavic out-migration from the non-Slavic south has increased greatly, primarily as a result of rising fears of anti-outsider and particularly anti-Russian sentiments building in the region (Rybakovskiy and Tarasova, 1990). Skilled Russians and other non-indigens are leaving the region by the thousands not for lack of work, but because of the anti-outsider or nativistic violence evident in the events of Fergana, Tashkent, Dushanbe, Osh' Oblast, and Novy Uzhen. For example, non-indigens wishing to leave Tadzhikistan recently ranked 'fear of the recurrence of the events of February' (anti-Armenian violence in Dushanbe) first (20.2 per cent); 'the change in the linguistic situation' second (11.6 per cent), and 'deterioration in the material conditions of life' only third (10.5 per cent) (FBIS, 18 September 1990, p. 74). This is resulting in a 'brain drain' in the cities of Soviet Central Asia, even while unemployment among the relatively unskilled rural indigens is increasing dramatically.

In Soviet Central Asia and Kazakhstan, the

Table 3.1 *Nation as a per cent of total republic population, 1959–89*

Nation	Per cent 1959	1970	1979	1989	Percentage point change 1959–70	1970–9	1979–89
Russian	83.8	82.8	82.6	81.5	−1.0	−0.2	−1.1
Ukrainian	76.8	74.9	73.6	72.7	−1.9	−1.3	−0.9
Belorussian	81.1	81.0	79.4	77.9	−0.1	−1.6	−1.5
Uzbek	62.1	65.5	68.7	71.4	3.4	3.2	2.7
Kazakh	30.0	32.6	36.0	39.7	2.6	3.4	3.7
Kirgiz	40.5	43.8	47.9	52.4	3.3	4.1	4.5
Tadzhik	53.1	56.2	58.8	62.3	3.1	2.6	3.5
Turkmen	60.9	65.6	68.4	72.0	4.7	2.8	3.6
Georgian	64.3	66.8	68.8	70.1	2.5	2.0	1.3
Azeri	67.5	73.8	78.1	82.7	6.3	4.3	4.6
Armenian	88.0	88.6	89.7	93.3	0.6	1.1	3.6
Lithuanian	79.3	80.1	80.0	79.6	0.8	−0.1	−0.4
Latvian	62.0	56.8	53.7	52.0	−5.2	−3.1	−1.7
Estonian	74.6	68.2	64.7	61.5	−6.4	−3.5	−3.2
Moldavian	65.4	64.6	63.9	64.5	−0.8	−0.7	0.6

Source: for 1959 and 1970: TsSU SSSR, 1973, Vol. 4, pp. 12–15. For 1979: TsSU SSSR, 1984. For 1989: Soyuz, 1990, no. 32.
Note: The figures in columns 1 through 4 above are equal to the percentage of the total union republic population comprised by the titular nation. For example, in 1959 83.3 per cent of the population of the RSFSR was comprised by Russians.

Table 3.2 *Per cent of national population in home republic, 1959–89*

Nation	Per cent 1959	1970	1979	1989	Percentage point change 1959–70	1970–9	1979–89
Russian	85.8	83.5	82.6	82.6	−2.3	−0.9	0.0
Ukrainian	86.3	86.6	86.2	84.7	0.3	−0.4	−1.5
Belorussian	82.5	80.5	80.0	78.8	−2.0	−0.5	−1.2
Uzbek	83.8	84.0	84.9	84.7	0.2	0.9	−0.2
Kazakh	77.2	79.9	80.7	80.3	2.7	0.8	−0.4
Kirgiz	86.4	88.5	88.5	88.2	2.1	0.0	−0.3
Tadzhik	75.2	76.3	77.2	75.3	1.1	0.9	−1.9
Turkmen	92.2	92.9	93.3	93.0	0.7	0.4	−0.3
Georgian	96.6	96.5	96.1	95.1	−0.1	−0.4	−1.0
Azeri	84.9	86.2	86.0	85.7	1.3	−0.2	−0.3
Armenian	55.7	62.0	65.6	66.7	6.3	3.6	1.1
Lithuanian	92.5	94.1	95.1	95.3	1.6	1.0	0.2
Latvian	92.7	93.8	93.4	95.1	1.1	−0.4	1.7
Estonian	90.3	91.8	92.9	93.8	1.5	1.1	0.9
Moldavian	85.2	85.4	85.1	83.4	0.2	−0.3	−1.7

Sources: for 1959: TsSU SSSR, 1962, Vol. 16, pp. 209–10. For 1970: TsSU SSSR, 1973, Vol. 4, pp. 321–4. For 1979: TsSU SSSR, 1984. For 1989: *Soyuz*, 1990, no. 32.
Note: figures represent the per cent of the nations' population living in their respective home republics. For example, in 1959 85.8 per cent of all Russians lived in the RSFSR.

in-migration from the north was more than offset by a high rate of natural increase among the Turkic Muslims resident in the area, resulting in the growing demographic dominance of indigens throughout the post-war period. As was noted above, the increasing proportion of each republic's population comprised by indigens was also enhanced during the 1980s by Russian and other non-indigenous (Meskhetian Turks) out-migration. At the same time, the percentage of each nation's population in its respective home republic declined for the first time in the 1980s (see Table 3.2), and this appears to represent the beginnings of a trend toward out-migration of the Turkic Muslims that has been predicted for some time now (Lewis et al., 1976). Nonetheless, at present it is a relatively weak trend, except perhaps in the case of the Tadzhiks, and must be treated with extreme caution. Indeed, given the rapidly rising level of unemployment and underemployment in Soviet Central Asia and Kazakhstan, surprisingly little inter-homeland migration has occurred to date.

In Transcaucasia, a combination of relatively high rates of natural increase, relatively low levels of non-indigenous in-migration, and also return migration to the home republic by Armenians and Azerbaydzhanis served to increase the demographic dominance of each indigenous nation in its home republic. The latter trend was especially intense during the recent violence over the status of Nagorno Karabakh (Goskomstat SSSR, 1988, pp. 61-70). However, while Armenians fled Azerbaydzhan by the thousands, their number increased in Nagorno Karabakh autonomous oblast (Goskomstat SSSR, 1989, p. 80). Clearly, the rising level of international conflict in Azerbaydzhan and the lack of support from Moscow have not diminished Armenian resolve to gain control over this segment of the perceived ancestral homeland.

Finally, in Moldavia a trend toward greater demographic internationalization during the 1960s and 1970s was reversed during the 1980s, though this was not apparently due to Russian or Ukrainian out-migration or Moldavian in-migration (TsSU SSSR, 1984; Soyuz, 1990, 32). On the contrary, while the percentage of Moldavians in Moldavia increased between 1959 and 1970, it decreased during the subsequent two intercensal periods, indicating that there was a net out-migration among Moldavians (see Table 2). However, the Jewish population in the republic declined precipitously during the 1980s as emigration to Israel increased, and this, together with the slow growth and even relative decline among the Slavic national communities shifted the demographic balance in favour of Moldavians.

As in most other republics, since the 1989 census there has been a dramatic worsening of international relations in Moldavia. As Moldavians have pressured non-Moldavians, including the indigenous Gagauz, to learn the Moldavian language and have in other ways acted in what non-Moldavians perceive as national chauvinism, the non-Moldavians have not reacted primarily by leaving the republic, although some out-migration of Russians and others is undoubtedly occurring. Rather, the Gagauz and Russian-speaking population have each declared their sovereignty on territory claimed as their ancestral homeland. Each of these proclamations has been rejected by the central republican authorities, creating a potentially explosive situation.

Overall, even though the level of international interaction has increased in the Soviet Union over time, most nations today, including the Russians, have remained highly concentrated in their respective home republics. A number of reasons may be suggested for this. Firstly, ethno-cultural benefits accrue to members of the indigenous nation, but are in general unavailable outside. Secondly, socioeconomic and political benefits have also been gained by the indigenous nations over time. Upward mobility in the socioeconomic and political sectors has been facilitated for indigens by a series of affirmative-action-type programs. With the exception of the dominant Russians, non-indigens have found it increasingly difficult to compete with members of the indigenous nations. Thirdly, there has been an increase in anti-'outsider' sentiments expressed by indigens as they become more geographically and socially mobilized, and consequently come into greater contact and competition with non-indigens for élite positions. This rise in nativism with development can be seen as an impediment to inter-homeland migration which has become much more intense (or at least much more visible) during the past few years. This tendency for rising international interaction to serve as a catalyst for activating national territoriality among indigens in the Soviet Union is similar to the nativistic behaviour found among India's 'sons of the soil' (Weiner, 1978) and in other multinational, multi-homeland states (Horowitz, 1985; Connor, 1986).

During a period of heightened nationalism that is expressed as nativism, we should expect ever greater concentrations of indigens in their respective home republics. This has become most evident in the Transcaucasian republics: not only have Armenians and Azerbaydzhanis returned to their respective home republics by the thousand, but Russians and others have also left the region and are going home. While reports of a net out-migration of Russians from Latvia (Sheehy, 1989, p. 8) appear incorrect, the level of net in-migration has decreased dramatically in the past year, declining from 11,200 in 1988 to just 2,300 in 1989 (Latvija, 1990, p. 35). While still too early to be treated as a definitive trend, this dramatic downturn in net in-migration may be seen as a preliminary sign that the negative 'push factor' of rising indigenous nationalism has begun to outweigh the positive 'pull factor' of higher living standards that had drawn non-indigens to the Baltic region.

Those non-indigens who are returning home are often much more nationalistic after their experiences with indigenous nativism. For example, Azerbaydzhanis returning to Azerbaydzhan from Armenia since 1987 have been found at the forefront of nationalist organizations, anti-Armenian demonstrations and acts of violence in Baku and elsewhere. The same is likely to be true of Russians leaving the Central Asian, Baltic or Transcaucasian republics for 'Mother Russia'. It is certain that more intense international contact has not bred better international understanding in the Soviet Union, let alone the creation of a new Soviet people.

We should expect that this demographic reconcentration will continue, and as it does the potential for international integration will also decline. The very belief that heightened assimilation would result from greater inter-homeland migration in the Soviet Union has been shown to be highly questionable. Indeed, the opposite is more likely to be the case – that greater international mixing in a multi-homeland environment (where there is no 'neutral turf') has generally stimulated nationalist sentiments among the indigenous populations, and the non-indigens subjected to this nativism have also often experienced rising nationalism as a result.

Cultural indigenization

One of the principal ways that this nativism has been expressed is in the cultural sphere. As mentioned above, one factor helping to explain the relative immobility of peoples across homeland borders is the fact that ethno-cultural benefits accrue to members of the indigenous nation, but not normally to individuals living outside their respective home republics. The only true exception to this general rule is that the Russian language and culture has in the past occupied a position of dominance throughout the Soviet Union. This Russian cultural dominance has increasingly been challenged by non-Russian indigens asserting that their own languages and cultures should occupy a dominant position within their own homelands. Many nationalists have declared that their nations are under attack due to the status of the Russian language and culture (Rukh, 1989, pp. 9–10). Nations whose homelands are recognized as union republics, and a number of those at the autonomous level within the federal hierarchy, have drafted and/or adopted legislation declaring their languages to be the lingua franca of the home republic (JPRS, 1989).

For most nations, the real threat of linguistic assimilation is not to members living within their own home republics, but to those living outside (see Table 3.3). In addition to Russification, these non-indigens have over time been pressured to learn the language of the indigenous nation, and this pressure has grown enormously over the past few years. The attempt on the part of indigens to promote their cultural attributes to a position of dominance is one factor likely to cause a reduction in the level of inter-homeland migration. As noted above, the shift in favour of the Tadzhik language was cited as the second factor in non-indigenous desires to leave the republic. The pressure on non-indigens to acculturate – either to Russian or to the indigenous national culture – has led to some national front organizations including in their

Table 3.3 *Native language retention by nation and location, 1979–89 (Per cent of nation claiming the language of the nation as the native language)*

Nation	% in Home Republic 1979	1989	% pt chg*	% outside Home Republic 1979	1989	% pt chg*
Russian	100.0	100.0	0.0	99.2	98.9	−0.3
Ukrainian	89.1	87.7	−1.4	43.8	44.6	0.8
Belorussian	83.5	82.2	−3.3	36.8	36.5	−0.3
Uzbek	98.8	98.7	−0.1	96.9	96.1	−0.8
Kazakh	98.6	98.6	0.0	92.8	90.5	−2.3
Kirgiz	99.6	99.5	−0.1	84.8	85.3	0.5
Tadzhik	99.3	99.3	0.0	92.8	92.9	0.1
Turkmen	99.2	99.2	0.0	91.1	89.4	−1.7
Georgian	99.5	99.7	0.2	67.3	68.9	1.6
Azeri	98.7	99.1	0.4	92.6	88.8	−3.8
Armenian	99.4	99.6	0.2	73.9	75.6	1.7
Lithuanian	99.7	99.6	−0.1	63.6	59.2	−4.4
Latvian	97.8	97.4	−0.4	55.3	44.3	−11.0
Estonian	99.0	98.9	−0.1	47.1	44.3	−2.8
Moldavian	96.5	95.4	−1.1	74.3	72.8	1.5

Source: for 1979: TsSU SSSR, pp. 71–137. For 1989: Goskomstat SSSR, 1989.
% pt chg* is equal to the percentage point change

programs the goal of providing cultural, linguistic and other support to members living outside the home republic (Rukh, 1989).

The recent dramatic increase in pressure on all non-indigens to learn the indigenous language has posed a problem for members of minority nationalities who are indigenous, but have been given either restricted autonomous status or no autonomy at all. Charges of national chauvinism have been lodged against Azerbaydzhan by Armenians living in Nagorno Karabakh, against Georgians by Abkhazians and Ossetians, against Moldavians by Gagauz and against Lithuanians by Poles. Clearly, Russification is not the only integrative pressure that has been placed on national communities in the state. This acculturation pressure, at least as perceived by these minority indigens, has resulted in increasing calls for sovereignty among them, which has in turn led to a more hostile reaction on the part of the Moldavians, Georgians and Azerbaydzhanis.

Beyond language, there has been a resurgence of interest in the history, religion, customs and general culture of the nations within the Soviet Union. Much more scholarly effort on the part of indigens is today focused on these elements of national rebirth or revival. This is to be expected, given the fact that in the past these areas of interest were severely circumscribed by the state. Also, the indigenous intelligentsias have tended to be concentrated in the cultural sectors (Arutyunyan and Bromley, 1986; Smith, 1988), and this has undoubtedly facilitated the process of cultural indigenization. However, while 'cultural autonomy' has been allowed as an element in the centre's nationality policies, cultural indigenization has gone further to become a means through which members of the indigenous nation are attempting to attain a position of primacy in their homelands. The promotion of the indigen's cultural attributes to a position of dominance may also be seen as a way that indigens can gain a competitive edge over outsiders in the social, economic and political spheres of life, which in turn would discourage non-indigens from migrating to the nation's homeland in the first place.

Social mobilization and indigenization

Beyond the cultural dimension, a positive

correspondence between social mobilization and indigenization is also evident in the Soviet Union, though this runs counter both to Marxism-Leninism and also to much Western theory regarding the relationship between modernization and national self-consciousness. According to the Leninist thesis on the national question, socioeconomic development, coupled with equalization, were the critical processes that, along with political socialization, would lead to the creation of a new Soviet people. Turn-of-the-century New York City, appearing as a melting pot into which new immigrants poured and out of which Americans emerged, had a strong impact on Lenin's thinking in this regard (Lenin, 1913). Rapid development, and with it urbanization, would similarly serve to fuse the ethnic communities of the Russian Empire into one unified citizenry. This belief has served as a fundamental principle in the construction of solutions to the national question ever since.

Without a doubt, the Soviet Union has undergone one of the most rapid rates of modernization and urbanization witnessed by the human race (Lewis and Rowland 1979, pp. 158–60). Along with this rapid development, there have been efforts to equalize socioeconomic levels of development across nations, though certain regions such as Central Asia continue to lag far behind the more advanced European areas (see Chapter Two). Nonetheless, socioeconomic development and equalization, if not equality, were occurring throughout much of the Soviet period. To take one measure of social mobilization, the rate of educational attainment among the major nations of the state has shown a great deal of convergence over the past seventy years, and particularly in the post-war period. In 1926 literacy rates ranged from a high of 74.5 per cent among Latvians to a low of only 2.2 per cent among Tadzhiks, with Russians occupying an intermediate position with 45.1 per cent (TsSU SSSR, 1929, vol. 17, pp. 8–15). By 1985, at least 60 per cent of each nation's population had attained a secondary education, and over time the difference between nations with the highest and lowest levels of education has lessened dramatically (see Table 3.4).

If the social mobilisation of the various national communities, along with some degree of equalization, held the key to international integration or assimilation into one Soviet people, the Soviet Union today should not be facing a serious national problem. Unfortunately for Gorbachev, the thesis described above has proven incorrect, not only in the Soviet Union but in most developed multinational, multi-homeland states. Soviet sociological and ethnographic surveys during the past decade or so show that increasing social mobilization has coincided with rising national particularism (Arutyunyan and Bromley, 1986). The contemporary national problem is not a problem of a maladjusted and underdeveloped rural populace reacting to their lowly status, but rather a problem that is most serious in the urban centres among the élite members of society. As the riots in Alma Ata and the nationalist demonstrations from Kiev to Yerevan and Yakutsk indicate, indigenous college students in particular are at the forefront of demands for change in favour of national separatism. Beyond this, it has also become apparent that the most developed nations in the state (i.e. Estonians, Latvians, Georgians and Armenians), are also the most nationalistic, which is certainly not a predicted outcome of the assimilation with development or 'diffusionist' thesis.

The territorial dimension of nationalism is crucial in explaining the seemingly counter-intuitive relationship between national self-consciousness and social mobilization. It is an element that is present in a multi-homeland society such as the Soviet Union but absent from an immigrant society such as the United States. One of the reasons that Leninist nationality policies failed to predict accurately the relationship between national solidarity and socioeconomic development was that turn-of-the-century New York City, comprised primarily of voluntary immigrants wishing to become Americans, was an invalid model on which to construct a solution to the nationality question in the multi-homeland Soviet Union. As members of the indigenous nation become more socially and geographically mobile (i.e. more urbanized), they come into closer contact with non-indigens, and compete with them for the resources of the homeland, including high-status occupations. Rather than competing as individuals solely on the basis of achievement or merit-oriented criteria, upwardly mobile indigens are more likely to assert that their status as indigens itself should guarantee them a

Table 3.4 *Rate of educational attainment by nation, 1959–85[1] (/1000 population age 10+ with a higher or secondary [complete or incomplete] education)*

Nation	1959	1970	1979	1985	Percentage Point Change 1959–79	1970–79	1979–89
Russian	378	508	661	709	13.0	15.3	4.8
Ukrainian	353	476	614	673	12.3	13.8	5.9
Belorussian	311	438	591	650	12.7	15.3	5.9
Uzbek	311	412	615	683	10.1	20.3	6.8
Kazakh	268	390	592	678	12.2	20.2	8.6
Kirgiz	299	400	590	666	10.1	19.0	7.6
Tadzhik	299	390	565	643	9.1	17.5	7.8
Turkmen	363	430	597	677	6.7	16.7	8.0
Georgian	474	578	725	784	10.4	14.7	5.9
Azeri	360	424	635	725	6.4	11.1	9.0
Armenian	443	518	719	769	7.5	20.1	5.0
Lithuanian	208	353	539	614	14.5	18.6	7.5
Latvian	426	488	618	696	6.2	13.0	7.8
Estonian	358	462	587	661	10.4	12.5	7.4
Moldavian	212	338	534	609	12.6	20.4	7.5
Coefficient of Variation[2]	22.0	14.3	8.7	7.0			

Sources: for 1959–79: Zinchenko, 1984, p. 160. For 1985: *Statisticheskiye materialy*, 1986, p. 6.

Notes:
1. Education data from the 1989 census were not available by the end of 1990. The 1985 figures are from the 5 per cent sample survey of the population taken 1 January 1985.
2. Coefficient of variation is one way to measure whether or not convergence or equalization is occurring. The figures above are given as per cents: the lower the per cent the lower the degree of variation between national rates of education. A CV of 0 per cent would equal complete equality, i.e. that all nations have exactly the same rate of educational attainment. Although this type of complete equality had not occurred by 1985, equalization according to this measure of social mobilization was clearly occurring between 1959 and 1985.

competitive advantage over 'foreigners'. This geographically distinct 'competition model' (Nielsen, 1985; Nagel and Olzak, 1982; Tiryakian and Nevitte, 1985), which is structured around the governing principle of national territoriality, is of paramount importance in understanding the failure of Soviet authorities to solve the national question.

Over time, as indigens have become more socially mobilized, they have been relatively successful in gaining greater control over élite positions in the socio-cultural (Smith, 1988), and increasingly in the economic and political sectors, of their respective homelands. Ironically, indigens have been assisted in this effort by the enactment of affirmative-action-type assistance targeting the indigenous nations for preferential treatment. These programs, part of the international equalization effort designed to overcome problems inherited from the past, have allowed for easier indigenous access to slots in institutions of higher education, and easier upward mobility within the governmental structure of the republic, in the Communist Party and also in economic enterprises (Karklins, 1984). However, these programs have been used by indigens in an effort to gain a dominant status within their respective homelands, and this has in turn raised the level of international tensions among the non-indigens who resent these programs (Rybakovskiy and Tarasova, 1990). A point that has been misunderstood in the Soviet Union and elsewhere is that upwardly mobile indigens ultimately feel they should be dominant in their own homelands; they are unlikely to be satisfied as mere equals with non-indigenous outsiders.

Successful indigenous social mobility has resulted in the creation of a cadre of educated nationalists in élite positions who are able to reorient the cultural, social, economic and political institutions of the home republic toward support for national self-determination. This is not equally true for every national community in the state, or even for all those whose homelands are recognized as union republics. For example, the nations of Soviet Central Asia, while becoming increasingly educated and socially mobilized in recent years, nonetheless remain predominantly rural and currently lack the kind of internal cohesion and socioeconomic and political control within their respective republics necessary to mobilize for greater national self-determination. This is not the only reason for a relative lack of overt support for national self-determination in Soviet Central Asia today. Their geopolitical position as a landlocked Asian region, their lack of a recent history of independence as the Baltic nations have experienced and their relative economic underdevelopment are additional factors helping to explain their relatively limited political objectives at present. This is not to suggest that the development of a more separatist national agenda cannot succeed in this region in the relatively near future. Four of the five Central Asian republics have adopted declarations of sovereignty that reflect an underlying strategy of national territoriality, i.e. an attempt to take control of the destiny of the nation by gaining sovereignty in the home republic. The fifth Central Asian republic, Kirgiziya, has drafted a declaration which appears much more international in its orientation (FBIS, 14 August 1990). This is undoubtedly a reflection not only of the high proportion of non-indigens living in the republic, but also an attempt to counter the rising fear of anti-outsider nativism after the events of Osh' Oblast, which has led to a dramatic increase in the number of skilled out-migrants from the republic (FBIS, 27 August 1990, pp. 117–18).

Political separatism

Much of the preceding discussion indicates that a process of national separatism, defined broadly as the indigenization of life in the national homeland, has represented a relatively long-term challenge to Soviet federalism. It could be argued that this national separatism manifested itself primarily in the socio-cultural arena in the past because this was politically acceptable in the Soviet Union. Before 1985, while cultural and social indigenization were to a great extent accommodated by Moscow, coercion or the threat of coercion kept outright calls for political separatism off the nationalist agenda. What is new about the national question since Gorbachev, then, is that this threat of coercion has diminished and national separatism has become overtly political. This should be seen as a new stage in a much longer process, rather than an entirely new dilemma facing the Kremlin.

Declarations of sovereignty

Nearly every indigenous national community, from Lithuanians to the Gagauz, Nogays and Chukchi are pressuring Moscow for greater political control over the territories perceived by each national membership as its rightful homeland. At this fundamental level, the national communities in the Soviet Union are on the same track. However, not every political separatist movement is equally threatening to the stability of the Soviet Union. The most serious challenge to Soviet federalism obviously comes from those nations that have been successfully mobilized behind the goal of secession and the creation of their own sovereign nation-states. The Lithuanian declaration of independence, and also the independence movements in Estonia, Latvia and Georgia currently fall into this category. These republics are unwilling to participate in the negotiation of a new union treaty, and unlike the other republics their declarations of independence are not proclaimed to be the basis for the creation of a new federation, but rather for the establishment of separate nation-states. The Moldavian and Armenian declarations appear close to this more secessionist-leaning stance, though each has held out the prospect of union treaty renewal, albeit on a much more confederal basis. There is also an interstate dimension to the political separatism of Moldavians (as well as Azerbaydzhanis); the relationship of Moldavians to Rumania and Azerbaydzhanis to their co-nationals in Northern Iran will undoubtedly influence decisions regarding secession from the Soviet Union. The support for secession among Armenians also appears tempered by the interstate ethno-political environment. Most Armenian supporters of independence have also recognized the need to 'normalize' relations with their Turkic Muslim neighbours before secession. However, at the same time the lack of support from Moscow over the conflict with Azerbaydzhan has weakened the argument that participation in a new Soviet Federation is essential for the preservation of Armenia (FBIS, 26 October 1990, pp. 88–9).

In Ukraine, Belorussia and Azerbaydzhan the national front organizations are opposed to any participation in a new union treaty, and have lobbied for declarations of independence, rather than declarations of sovereignty within a renewed Soviet federation. However, the national fronts in all three republics were unsuccessful in winning a majority or plurality in the republican Supreme Soviets, and so the process of negotiating a new federal relationship is proceeding at present in these republics. The support for these national fronts, and Rukh in particular, has increased substantially since the elections. Should the national front organizations in these three republics succeed in gaining a majority in the next elections, a likely prospect if fair elections are held, they will almost certainly join the other western nations in declaring their independence from the Soviet Union. At present, each is placing pressure on the centre to confederalize the relationship between itself and the republics. While less threatening to the future viability of the Soviet Union than secessionist movements, this confederalist pressure nonetheless represents a serious challenge to the future of the Soviet state. If successful, confederalization of the state would effectively end Moscow's independent political dominance and could easily lead to the dissolution of the state as well, albeit at a less frenzied pace. This would be comparable to the situation in Yugoslavia, where political and economic reforms converted the state into a confederation of essentially sovereign nation-states by the mid-1970s. As anyone familiar with Yugoslavia knows only too well, this confederalization did not solve the national problem in the state, which is currently facing independence movements in Slovenia and Croatia, and violent international conflict in Kosovo between Albanians and Serbs over the question of sovereignty.

The Russian Republic under Yeltsin's leadership also appears to favour greater decentralization than mere 'federalism', but there is a great diversity of opinion in the republic. The December 1990 attempt to reduce the amount that the republic contributes to the federal budget is just one aspect of the confederal assertiveness evident in the RSFSR. The republic has also concluded a number of bilateral treaties and trade agreements with other union republics, bypassing the central authorities and perhaps obviating the need for a new union treaty, though Yeltsin continues to maintain his support for such a document (*Moscow News*, 1990, 48, p. 5). In the projected Constitution of the Russian Federation, the republic proclaims the right of a sovereign state to control its own resources and economy, conduct its own domestic and foreign affairs, and protect itself from internal and external threats (*Argumenty i Fakty*, 1990, 47, p. 3). In other words, there appear to be few functions left for an inter-republican central authority to perform. By the end of 1990 the final version of the constitution and the position of the leadership regarding a new union treaty were still very much in flux.

At a lower geographic scale, political decentralization is occurring within the RSFSR as the federation treaty is renegotiated. Nearly every autonomous unit within the RSFSR has declared its sovereignty, albeit within a renewed Russian Federation. The draft constitution, along with the declaration of sovereignty, recognizes the multinational, multi-homeland nature of the RSFSR and does not provide any special status for the Russian nation, except for the recognition of Russian as the republic-wide language. In addition, those national communities whose homelands were recognized at different levels in a federal hierarchy are now envisioned as equal republics in the new federation, a definite improvement in status for the former 'autonomies' (*Argumenty i Fakty*, 1990, 47). These moves, along with the declarations of so-called 'free enterprise zones' made by several non-autonomous oblasts in the republic, have been denounced as attempts to destroy the Russian nation and dismember its homeland (FBIS, 12 September 1990b, p. 90). Combined with those Russian nationalists who view the homeland as something more than the republic, this opposition may become a potent force for continued central authority.

At present Russian nationalism in the RSFSR is still not a cohesive movement, and it is too early to tell whether it will work toward the renewal of Soviet federalism or to undermine it further (Szporluk, 1989). Outside the RSFSR, Russian nationalists have tended to see themselves as the loyal opposition to indigenous separatism, and may be a force that the central authorities can count on to counteract the divisive nature of contemporary indigenous nationalism in the non-Russian periphery. However, attitudes toward secession among these outlying Russians vary by region (Gudkov, 1990). The treatment of these Russians living outside the RSFSR has also served as a catalyst in the activation of Russian nationalists within the RSFSR. For example, they have been successful in forcing a delay of the signing of a bilateral treaty between the RSFSR and Estonia over the proposed exclusion of Russians and others who entered Estonia after 1940 from taking part in elections (FBIS, 26 October 1990, p. 82).

Finally, Soviet Central Asia, with its rapidly growing population and its declining economic conditions, along with its adherence to Islam at a time of rising Islamic fundamentalism across the border in Iran and Afghanistan, was proclaimed the region that would cause the disintegration of the Soviet Union (Bennigsen, 1979; Carrere d'Encausse, 1978). Interestingly enough, this is the one region that poses almost no immediate threat to the political structure of the state. The primary reason that a number of Western analysts missed so badly in predicting that Central Asia would be the Achilles heel of the Soviet Union was the mistaken assumption that all of the communities whose religion was Islam comprised one Muslim nation united in its opposition to the atheistic Soviet Union (Bennigsen, 1979). However, as recent events in Fergana and Osh' Oblast have made abundantly clear, the Turkic-Muslims in the Soviet Union are divisible into numerous nations, each with its own political agenda.

The only coherent national-front organization in the region is *Birlik* ('Unity') in Uzbekistan; its goals are oriented at present more toward enhancing the socio-cultural status of the language and the diversification of the economy away from cotton and greater political autonomy within the Soviet Union than toward political independence for the republic or the region (*Uzbek Popular Front*, 1989, pp. 4–7). In a recent survey of public opinion on a wide range of issues, Central Asians responded much more conservatively than did Georgians or members of the Baltic nations, indicating that the political mobilization of indigens behind the goal of national self-determination is not as advanced in this region of the country (Mickiewicz, 1990). The declarations of sovereignty made by these union republics also appear to be more supportive of a federal arrangement with a continued strong centre. Ironically, given the growing socioeconomic problems and the ethnocultural differences of Soviet Central Asia, this is the one region that the centre may be willing to give up without too much of a struggle.

The support for a renewed Soviet federation in the region does not mean that all is well in Soviet Central Asia. Given the declining socioeconomic conditions and the perception among indigens that only they should rightfully benefit from the development that is occurring, rumours that outsiders have been given preference in hiring or in the allocation of other resources such as housing, will continue to result in a violent nativistic reaction. This is likely to increase in intensity as conditions worsen. However, this type of international conflict, while serious, is not of the same order of threat to the viability of the Soviet Union as is the pressure for confederalization or outright secession.

The question of 'autonomization' revisited

Given the theme of this chapter – nationalism as a challenge to Soviet federalism – we have thus far focused our attention on the nations whose homelands are recognized as sovereign union republics with the right to self-determination, up to and including secession. However, it is necessary briefly to examine the trends in political separatism among those nations whose homelands are designated as 'autonomous' within the federal hierarchy.

Below the union republic level nations, there are a growing number of challenges to Soviet federalism as it was structured in the past. The first is the effort by nations whose homelands are recognized as autonomous republics to have that status upgraded to union republic. Only union republics have the constitutional right to self-determination. For autonomous republics political separatism is at least a two-stage process: the first is the attainment of union republic status, the second the confederalization of political power relations with the centre. The Tatars, Bashkirs, Yakuts and Abkhazians have been the most vocal in this regard, but almost every autonomous republic, and even the autonomous oblasts and okrugs in the RSFSR, have declared their 'sovereignty'. However, aside from the Tatar declaration, which proclaimed the Republic of Tatarstan to be independent from the Russian Federation (FBIS, 5 September 1990, pp. 115–16), these documents proclaim the continued desire to remain a part of the RSFSR in a renewed Soviet Union.

In addition to the expressed desire to gain greater socio-cultural and economic sovereignty (particularly control over natural resources), these declarations should be seen as an attempt to enhance the bargaining position of the autonomous units as the new union treaty and the new Russian federation treaty are negotiated. These declarations are motivated in large part by the failure of the central authorities to treat the autonomous republics as equals of the union republics in each of these two political processes, as well as in the Shatalin economic plan, which recognized the union republics as sovereign, but said nothing about the autonomies. In these cases, localists of all nationalities have found their interests coinciding with those of the indigenous nationalists seeking greater control over the fate of their nations. However, while these declarations of sovereignty favour further decentralization, it is not at least immediately evident that they favour the disintegration of the Soviet Union. Indeed, quite the opposite is likely to be the case, since many of these national communities will continue to look toward a strong centre to offset the 'national chauvinism' of the locally dominant union-republic nations.

The attempts by these national communities to raise the status of their homelands to the level of 'sovereign' union republics is likely to bring them into conflict with the nations in whose home republic they are currently incorporated (see Map 3.1). This has already occurred in Georgia, where a petition by Abkhazians to 'secede' and form their own union republic and a declaration of sovereignty by the South Ossetian Autonomous Oblast have resulted in a violent confrontation with Georgian nationalists who viewed these efforts as attempts to remove pieces of the Georgian homeland from its rightful owners. A similar nationalistic reaction may be expected from at least some elements of the Russian nation, should the Tatars or other national communities in the RSFSR be successful in gaining their independence from Russia.

A number of indigenous national communities in the state whose homelands are not recognized in the federal structure at all are petitioning for such recognition today. The first group are those nations who were deported from their homelands during and immediately after the Second World War for alleged collaboration with the Nazis. The Ingush and Karachay, which had their own autonomous republics before the Second World War but after 1956 were given only shared status in the Chechen-Ingush ASSR and the Karachay-Cherkess Autonomous Oblast, have proclaimed their independence in new sovereign republics. The Germans and Crimean Tatars, who had autonomous republics before the Second World War, are also petitioning for the recreation of these units. Possible sites for a new German 'homeland' in the Soviet Union include the location of the former German republic in the lower Volga region and the current area of German concentration in northern Kazakhstan. Some have mentioned Kaliningrad Oblast as a possibility. This site was raised with an eye to interstate investments: a Kaliningrad German region is envisioned by some as a possible high technology showcase to attract West German capital to the Soviet Union (*Nemetskaya avtonomiya*, 1989, p. 11). Crimean Tatars have been given permission to resettle in Crimea, and have flooded there as a result. Over

100,000 of an estimated 350,000 total have already moved back to Crimea; this massive return migration stream has led to serious international conflict in the region (Kiselev, 1990). The level of return migration after forty-five years in exile is dramatic testimony to the power of the homeland idea and was clearly unexpected by the centre. A referendum on the status of Crimea, in which Crimean Tatars throughout the country were allowed to participate, took place on 20 January 1990.

Aside from the attempts by nationalists from the deported nations to recover their lost homelands, the demand for official recognition of homeland status is also being made by national communities that are minorities in republics where a push for secession by the dominant nation has either already happened or is likely in the near future. As we noted above, the Gagauz and Russian-speaking population concentrated on the left bank of the Dnestr River have declared their sovereignty in Moldavia; Poles in Lithuania have petitioned for similar autonomy. For the Poles in Lithuania or the Gagauz and Russian-speaking population in Moldavia, the desire for autonomous republics of their own may reflect the minorities' concern with what life would be like in an independent Lithuania or Moldavia. Since the law on secession allows autonomous republics to take a separate vote for or against secession if such a referendum on independence is held in the republic, these national communities appear to be positioning themselves to opt out of Moldavia and Lithuania should these two republics follow the procedures in the law on secession. In the case of these minority national communities, paradoxically, a move for greater sovereignty may also be a move in support of continued participation in a new Soviet Union and not in favour of its demise. The same is undoubtedly true in the case of Abkhazia and South Ossetia.

Central responses

Soviet federalism as an instrument designed to help overcome past international antagonisms and thereby help to create a united Soviet people has proved a dismal failure. The movement toward one Soviet people, or even toward one Russian nation, has been extremely limited during the past seventy years. On the contrary, national separatism in its broadest sense (indigenization) has represented a stronger force for devolution during the entire post-war period. This national separatism has entered a new overtly political phase since 1985. In this latest stage, Moscow has increasingly found itself in a reactive position as nationalists from around the country are setting their own political agendas and mobilizing their national memberships behind these goals. The need to resort to the use of military force both to intimidate nationalists and to restore order in Lithuania, Tbilisi, Baku, Dushanbe, Tashkent, Kirgiziya, Nagorno Karabakh and Moldavia is itself a strong indication that Moscow has lost the initiative in managing the national question. Beyond this, events such as the 9 April 1989 confrontation in Tbilisi have served as catalysts to mobilize the nation behind political separatism; in this sense they have proven extremely counterproductive. In addition to threatened or actual use of coercion to shore up the centre's authority, the political leadership in the Soviet Union is attempting to regain control over the national problem through legislative means. The primary laws designed to accomplish this are the law on secession and the union treaty.

The law on secession

According to Goble (1990, p. 1) the 'new law on secession is intended not to facilitate independence for the republics but to make it impossible'. However, compared with the situation in previous years, when even espousing the goal of secession was enough to justify imprisonment, the current situation is obviously much more accommodative of the desire for self-determination. The very need to have such a law on the books is a strong indication that the political leadership is no longer able to keep the issue of national self-determination off the political agenda. While this right has been enshrined in the Soviet Constitution since the founding of the state, the law on secession is profound in the sense that it is the first legal mechanism through which the union republics may exercise their constitutional rights in this regard (*Pravda*, 1990, p. 2).

Nevertheless, there are problems with it from a nationalistic point of view. The most serious is that it is not a law recognizing the right to *national* secession, but rather the right of union republics (and only union republics) to secede from the Soviet Union. The new law on secession calls for a popular referendum that must be approved by two thirds of the eligible voters in the union republic (*Pravda*, 1990, p. 2). For a number of nations that do not have a demographic preponderance in their home republics (see Table 3.1), the difference may be critical, since it is almost certain that national separatist movements will be more strongly supported by indigens than non-indigens. For example, while 91 per cent of indigens in the Baltic republics favour secession, only 41 per cent of the Russians living there are for independence from the Soviet Union (Gudkov, 1990). In an effort to circumvent this problem, there has been a move by Estonian nationalists to limit the population eligible to vote in a referendum on secession to the pre-1940 citizens of Estonia and their descendants (i.e. the population of independent Estonia). Similar restrictions on republican citizenship have been proposed in Latvia, Lithuania, Moldavia and Georgia. In addition, since the three Baltic nations, along with Georgia, were independent earlier in the century and were forcibly reincorporated into the Soviet Union, none of them recognizes the validity of the law itself. The 'Republic of Moldova', the majority of which was incorporated into the Soviet Union after the Second World War, is also likely to use this argument in an effort to circumvent the law on secession. Historic justification rather than the acceptance of a popular referendum is the basis on which the Baltic nations have declared their independence from the Soviet Union.

The purpose of the law on secession should not be viewed as facilitating the disintegration of the Soviet Union into fifteen independent nation-states, but rather as a legal mechanism through which the centre can regain control over the pace of political separatism in the country. The main goal of the law appears to be to discourage secession as strongly as possible, while making the act itself legal. This is understandable from Moscow's point of view. The question remains as to whether it will make it so difficult to secede that nationalists who are determined to gain independence will seek change through revolutionary means.

Negotiating a new Union Treaty

The other major legal mechanism is the new Union Treaty, designed to restructure the federal relationship between centre and republic. While in principle the creation of a Union Treaty is the responsibility of the republics themselves, the central leadership has proposed draft versions of this document in an effort to set the political agenda. The draft Union Treaty endorsed by Gorbachev was published before the December meeting of the Congress of Peoples Deputies (Gorbachev 1990) which endorsed the concept of a Union Treaty. Public opinion polls showing majority support for a union treaty have also been published (FBIS, 24 December 1990, pp. 72–3). However, there has been a strong negative response to this draft by several of the republican leaderships. The powers claimed for the centre in Gorbachev's proposed Union Treaty include the preservation of the sovereignty and territorial integrity of the entire Union (i.e. control over the Soviet armed forces); the representation of the Soviet Union in foreign political and economic affairs, and the coordination of the foreign political and economic affairs of the republics; the development with the republics of a strategy for creating an all-union market, including the development of financial, credit and monetary policies, formation and utilization of a budget; joint centre-republican control over the energy, transportation and communication systems, the utilization of natural resources and the formation and implementation of ecological policies; the establishment with republics of social welfare policies for the population; coordination of inter-republican research and development in the areas of science, culture and education; and coordination of activities to preserve law and order (Gorbachev, 1990). For the most part, this list gives the appearance that the centre would function primarily to facilitate and coordinate the activities of the republics that would impinge on other republics or on the union's relations with other states. However, the way in which the centre and republics would share control is left extremely vague. The draft provides for a 'Soviet of the Federation' (Federation Council), which would essentially act as an advisory board to the president. This subordinate role is clearly unacceptable to the majority of republican leaders. Nevertheless, the contemporary Federation Council, which was given greater decision-making authority in December 1990, may serve in the future as an important forum in a renewed federation, if the president cedes real power to it, and if the republican governments allow their representatives to make decisions for them.

The centre's proposed control – whether jointly or independently – over the army, natural resources, foreign affairs and internal security brings the draft Union Treaty into direct conflict with the declarations of sovereignty made by several of the union republics, which proclaim the exclusive right to use their own natural resources, formulate their own interstate policies and defend and protect their own interests. Particularly unacceptable is the clause on protecting the 'territorial integrity of the Union', especially since the draft Union Treaty includes no mention of the right of the union republics to secede from the union. Indeed, the republics are not even said to have the right to self-determination, which would instead be vested in 'each people'. This seems further to confuse the issues of sovereignty, self determination and the potential utility of the law on secession. Perhaps the most problematic aspect of the draft Union Treaty is that it was crafted by the centre, reflecting 'old thinking' by Gorbachev and his colleagues. Since the 1922 Union Treaty and the Soviet constitutions provided for confederal relations in theory, nationalists in the Soviet Union today are likely to be quite sceptical of any new Union Treaty handed down from above.

At present, the union republics are engaged in concluding bilateral and multilateral treaties and trade agreements with one another, in order to establish new working relationships. This process may culminate in the signing of a new Union Treaty, in which the republics agree to cede to a central apparatus certain all-union functions. However, for some nations at least, even confederalization no longer appears acceptable. Lithuanians, Latvians, Estonians and Georgians (at minimum) have moved beyond a point at which confederation is a satisfactory political arrangement; at least at present they do not appear to be willing to accept anything less than secession and the creation of their own independent nation-states.

Conclusions

National territoriality, or the attempt by nationalists to gain control over the destiny of the nation by asserting their right to exclusive control of the national 'homeland', has resulted in a crisis that is certain to reshape the political geography of the Soviet Union. Short of military intervention and extreme coercive measures, a number of national communities will begin life as politically independent (though economically dependent) states in the near future. Those national communities remaining in a renewed Soviet federation are almost certain to have much greater control over their own lives in their own homelands. Whether this will be a loose federal arrangement or a confederal system remains to be seen. Political decentralization, if it is extensive enough to satisfy the national territoriality ambitions of the majority of the members of most nations in the state, should allow for the revitalization of Soviet federalism. However, this 'confederalization' of the state is certain to lead to new and perhaps more intractable national problems, which a weakened centre will be even less capable of managing.

Notes

1. The author gratefully acknowledges the support received from the Kennan Institute for Advanced Russian Study and the Center on East-West Trade, Investment and Communications while researching this topic.
2. For a discussion of the use of this model, and the need for a new contextual framework, see Lewin (1987) and Clem (1988). The totalitarian model continues to be favoured by a number of analysts. See, for example, Shtromas and Kaplan (eds), 1988.
3. It should be emphasized that Russification was only an active policy at the very end of the Tsarist period. Before then, the tsars generally promoted the idea that they represented all the peoples in the empire (Rossiya), or at least all those who practised the Orthodox faith, and not just the Russians and

Russiya (Kristof, 1967). Still, the Russian language and cultural values did serve as the core or dominant values around which the state developed, providing concrete relative advantages to Russians that continued after the October Revolution (Clem, 1988, pp. 41–3).

4. Other reasons for the reversal in favour of federalism included the utility of a federation of equal nations in propaganda directed toward the Third World colonies, and also the fact that federation at a time when the Russian Empire was disintegrating represented a move in the direction of reunification (Connor, 1984, pp. 51–7). It was ultimately acceptable because real political power was to be retained in the hierarchically structured communist party (Pipes, 1968).
5. Even as late as 1989, Gorbachev still appeared to believe that the Soviet population had for the most part been successfully socialized and that nationalism was the work of a few deviants. For example, at the September Party Plenum held to discuss the national problems facing the country, Gorbachev (1989, p. 13) stated that 'We have grown up in a social atmosphere literally permeated with internationalism. Friendship of the peoples was not some kind of abstract slogan for us, but an everyday reality. Can we really forget that? Can we renounce the internationalist legacy of our revolution?' One wonders to whom the 'we' in the above quote refers.
6. By the end of 1990, Kirgiziya had not yet adopted a declaration of sovereignty.
7. The issues related to economic indigenization, particularly the international equalization question and economic autarky, are examined here by Liebowitz (Chapter Two) and will not be covered here.
8. The internal passport and 'propiska' systems have had a dampening effect on movement (Zaslavsky and Luryi, 1979; Smith, 1985). However, according to Rybakovskiy and Tarasova (1989), contemporary migration problems arise from the fact that migrants respond primarily to larger societal forces and are not controlled by such instruments as the internal passport. This lack of control has become especially apparent since 1985.

Suggested reading

Smith, G. (ed.) (1990), *The Nationalities Question in the Soviet Union*, London, Longman.

References

Allworth, E. (1973), 'Regeneration in Central Asia', in *The nationality question in Soviet Central Asia*, Allworth, E. (ed.), New York, Praeger.

'Argumenty i fakty, 1990, Rossiyane: natsional'nyy sostav Rossii po dannym goskomstata', RSFSR, *Argumenty i fakty, (33), p. 1.*

'Argumenty i fakty, 1990, Konstitutsiya Rossiyskoy Federatsii (proyekt)', *Argumenty i fakty*, (47), special report.

Arutyunyan Yu., Bromley, Yu., eds., 1986, *Sotsial'no-kul'turnyy oblik sovetskikh natsiy*, Moskva, Nauka.

Bahry, D (1987), *Outside Moscow: power, politics, and budgetary policy in the Soviet republics*, New York, Columbia University Press.

Bennigsen, A. (1979), Several nations or one people? Ethnic consciousness among Soviet Central Asian muslims, *Survey* **24** (3), pp. 51–64.

Bromley, Yu. (1983), *Ocherki teorii etnosa*, Moskva, Nauka.

Bruk, S. (1986), *Naseleniye mira: etnodemograficheskiy spravochnik*, Moskva, Nauka.

Carrere d'Encausse, H. (1978), *Decline of an empire: the Soviet socialist republics in revolt*, New York, Harper & Row.

Clem, R. (1988), 'Perestroika: the social context of reform in the Soviet Union', in *Gorbachev's new thinking*, Liebowitz, R. (ed.), Cambridge, Mass., Ballinger.

Connor, W. (1984), *The national question in Marxist-Leninist theory and strategy*, Princeton, N.J., Princeton University Press.

Connor, W. (1986), 'The impact of homelands upon diasporas', in *Modern diasporas in international politics*, Sheffer, G. (ed.), London, Croom Helm.

Conquest, R. (1967), *Soviet nationalities policy in practice*, New York, Praeger.

Emerson, R. (1960), *From empire to nation: the rise to self-assertion of Asian and African peoples*, Boston, Beacon Press.

FBIS (14 June 1990), Text of sovereignty declaration (RSFSR) published, *FBIS-SOV-90-115* (Soviet Union: daily report), pp. 102–3.

FBIS (14 August 1990), Kirghiz declaration up for nationwide discussion, *FBIS-SOV-90-157* (Soviet Union: daily report), pp. 86–7.

FBIS (27 August 1990), Kirghizia's sovereignty future analysed, *FBIS-SOV-90-166* (Soviet Union: daily report), pp. 117–18.

FBIS (5 September 1990), 'Republic of Tatarstan' sovereignty detailed, *FBIS-SOV-90-172* (Soviet Union: daily report), pp. 115–16.

FBIS (12 September 1990a), Declaration of Turkmen sovereignty published, *FBIS-SOV-90-177* (Soviet Union: daily report), pp. 107–8.

FBIS (12 September 1990b), Isakov's views on union

treaty countered, *FBIS-SOV-90-177* (Soviet Union: daily report), pp. 89–92.

FBIS (18 September 1990), Tajikistan tension 'shows no sign of abating', *FBIS-SOV-90-181* (Soviet Union: daily report), pp. 73–4.

FBIS (28 September 1990), Tense ethnic situation reported in 'Moldova', *FBIS-SOV-90-1896* (Soviet Union: daily report), pp. 95–7.

FBIS (26 October 1990), Armenia's Ter-Petrosyan sets policy objectives, *FBIS-SOV-90-208* (Soviet Union: daily report), pp. 87–90.

FBIS (16 November 1990), Paper views intransigent attitudes, *FBIS-SOV-90-222* (Soviet Union: daily report), pp. 78–83.

FBIS (24 December 1990), Poll indicates majority support union, *FBIS-SOV-90-247* (Soviet Union: daily report), p. 72.

Gleason, G. (1990), *Federalism and nationalism: the struggle for republican rights in the USSR,* Boulder, CO, Westview Press.

Goble, P. (1990), 'Gorbachev, secession, and the fate of reform', *Report on the USSR* (Radio Liberty) 2 (17): pp. 1–2.

Gorbachev, M. (1988), Rech' General'nogo Sekretarya TsK KPSS M. S. Gorbacheva na plenume TsK KPSS 18 Fevralya 1988 goda, *Pravda,* 18 February 1988.

Gorbachev, M. (1989), O natsional'noy politike partii v sovremennykh usloviyakh, Moskva, Politizdat.

Gorbachev, M. (1990), O proyekte novogo soyuznogo dogovora, *Izvestiya,* 24 November, pp. 1–2.

Gorbatov, A. (1990), Bolshiye plany vokrug 'Maloy Litvy', *Pravda,* 22 March, p. 2.

Goskomstat SSSR (1988), Naseleniye, *Statisticheskiy press-bulleten',* (4), pp. 49–79.

Goskomstat SSSR (1989), *Natsional'nyy sostav naseleniya,* Moskva, Finansy i Statistika.

Gudkov, L. (1990), Russkiye v respublikakh, *Moskovskiye novosti,* (41), p. 7.

Guthier, S. (1977), 'The Belorussians: national identification and assimilation, 1897–1970', (Part I, 1897–1939), *Soviet studies,* 29 (1), pp. 37–61.

Hodnett, G. (1967), 'The debate over Soviet federalism', *Soviet studies,* 18 (4), pp. 458–81.

Horowitz, D. (1985), *Ethnic groups in conflict,* Berkeley, University of California Press.

JPRS (1989), 'Republic language legislation', *JPRS-UPA-89-063* (Soviet Union: political affairs), 5 December.

Kaiser, R. (1988), National territoriality in multinational, multi-homeland states: a comparative study of the Soviet Union, Yugoslavia and Czechoslovakia. Ph.D. dissertation, Columbia University.

Karklins, R. (1984), 'Ethnic politics and access to higher education', *Comparative politics,* April, pp. 277–94.

Kiselev, V. (1990), 'Deportirovannyye. Samozakhvatchiki. Chto dal'she?', *Moskovskiye novosti,* (45), pp. 8–9.

Kommunist (1961), Programma *kommunisticheskoy* partii Sovetskogo Soyuza, *Kommunist,* **38** (16), pp. 20–101.

Kristof, L. (1967), 'The geopolitical image of the fatherland: the case of Russia', *The western political quarterly,* pp. 941–54.

Latvija (1990), *Latvija skaitlos 1989 gada,* Riga, Latvija Republikas Valsts Statistikas Komiteja.

Lenin, V. (1913), 'Critical remarks on the national question', in *National liberation, socialism and imperialism,* pp. 12–44. Selected writings by V. Lenin, New York, International Publishers.

Lenin, V. (1914), 'On the national pride of the Great Russians', in *The Lenin anthology,* pp. 654–8, Tucker, R. (ed.), New York, W. W. Norton.

Lenin, V. (1922), 'The question of nationalities or "autonomisation"', in *The Lenin anthology,* pp. 719–24, Tucker, R. (ed.), New York, W. W. Norton.

Lewin, M. (1987), *The Gorbachev phenomenon: a historical interpretation,* Berkeley, University of California Press.

Lewis, R., Rowland, R., Clem, R. (1976), *Nationality and population change in Russia and the USSR,* New York, Praeger.

Lewis, R., Rowland, R. (1979), *Population redistribution in the USSR,* New York, Praeger.

Liebowitz, R. (1991), 'Inequality under Gorbachev', in *The Soviet Union: a new regional geography?,* Bradshaw, M. (ed.), London, Bellhaven Press.

Mickiewicz, E. (1990), 'Ethnicity and support: findings from a Soviet-American public opinion poll', *Journal of Soviet nationalities,* **1** (1), pp. 140–7.

Moscow news (1990), 'Russia up to its neck in work', *Moscow news,* (48), p. 5.

Nagel, J., Olzak, S. (1982), 'Ethnic mobilization in new and old states: an extension of the competition model', *Social problems,* **30** (2), pp. 127–43.

Nemetskaya avtonomiya, 1989, *Literaturnaya gazeta,* 11 October, p. 11.

Nielsen, F. (1985), 'Toward a theory of ethnic solidarity in modern societies', *American sociological review,* **50**, April, pp. 133–49.

Pearson, R. (1983), *National minorities in Eastern Europe 1848–1945,* London, Macmillan.

Petro, N. (1987), 'The project of the century: a case study of Russian national dissent', *Studies in comparative communism,* **20** (3/4), pp. 235–52.

Pipes, R. (1968), *The formation of the Soviet Union,* revised edn.), New York, Artheneum.

Pravda (1990), Zakon SSSR o poryadke resheniya

voprosov, svyazzanykh s vykhodom soyuznoy respubliki iz SSSR, *Pravda,* 7 April, p. 2.

Rakowska-Harmstone, T. (1970), *Russia and nationalism in Central Asis: the case of Tadzhikistan, Baltimore,* Johns Hopkins Press.

Rukh (1989), *The popular movement of Ukraine for restructuring (RUKH): program and charter,* Ellicott City, MD, Smoloskyp Publishers.

Rybakovskiy, L., Tarasova, N. (1989), Sovremennyye problemy migratsii naseleniya SSSR, *Istoriya SSSR,* (2), pp. 68–81.

Rybakovskiy, L., Tarasova, N. (1990), Migratsionnyye protsessy v SSSR: novyye yavleniya, *Sotsiologicheskiye issledovaniya,* (7), pp. 32–42.

Sack, R. (1986), *Human territoriality: its theory and history,* Cambridge, Cambridge University Press.

Schroeder, G. (1990), 'Nationalities and the Soviet economy', in *The nationalities factor in Soviet politics and society,* Hajda, L. and Beissinger, M. (eds), Boulder, CO, Westview Press.

Schwartz, L. (1986), Regional population redistribution and national homelands in the USSR. Ph.D. dissertation, Columbia University.

Sheehy, A. (1989), '1989 census data on internal migration in the USSR', *Report on the USSR* (Radio Liberty), 10 November, pp. 7–9.

Shevtsov, V. (1982), *The state and nations in the USSR,* Moscow, Progress Publishers.

Shtromas, A., Kaplan, M. (eds) (1988), *The Soviet Union & the challenge of the future,* Vol. 1, New York, Paragon House Publishers.

Smith, A. (1981), 'States and homelands: the social and geopolitical implications of national territory', *Millennium: journal of international studies,* 10 (3), pp. 187–202.

Smith, A. (1986), *The ethnic origins of nations,* London, Basil Blackwell.

Smith, G. (1985), 'Ethnic nationalism in the Soviet Union: territory, cleavage, and control', *Environment and planning c : government and policy,* (3), pp. 49–73.

Smith, G. (1988), 'Ethnoregional societies, "developed socialism" and the Soviet ethnic intelligentsia', in *Nationalism, self-determination and political geography,* Johnston, R., Knight, D., and Kofman, E. (eds), London, Croom Helm, pp. 166–88.

Soja, E. (1971), *The political organization of space* (resource paper 8), Washington, D.C., Association of American Geographers.

Solzhenitsyn, A. (1990), Kak nam obustroit' Rossiyu, Komsomolskaya pravda, 18 September, special report.

Soyuz (1990), Nas 285,761,976 chelovek, i vse my zhivem v odnoy strane, *Soyuz,* (32).

State (1990), Ethnicity and political boundaries in the Soviet Union, 0858 3–90 State (INR/GE), Washington D.C., US Department of State, Office of the Geographer.

Statisticheskiye materialy (1986), Itogi vyborochnogo sotsial'no-demograficheskogo obsledovaniya naseleniya 1985 goda, *Vestnik statistiki,* (7), pp. 67–80.

Szporluk, R. (1989), 'Dilemmas of Russian nationalism', *Problems of communism,* 38, July–August), pp. 15–35.

Szporluk, R. (1990), 'The imperial legacy and the Soviet nationalities', in *The nationalities factor in Soviet politics and society,* Hajda, L. and Beissinger, M. (eds), Boulder, CO, Westview Press.

Tiryakian, E., Nevitte, N. (1985), 'Nationalism and modernity', in *New nationalisms of the developed west,* Tiryakian, E. and Rogowski, R. (eds), Boston, Allen & Unwin, pp. 57–86.

TsSU SSSR (1929), *Vsesoyuznaya perepis' naseleniya 1926 goda,* 66 Vols., Moskva, TsSU SSSR.

TsSU SSSR (1962), *Itogi vesesoyuznoy perepisi naseleniya 1959 goda,* 16 Vols., Moskva Gosstatizdat.

TsSU SSSR (1973), *Itogi vsesoyuznoy perepisi naseleniya 1970 goda,* 7 Vols., Moskva, Statistika.

TsSU SSSR (1984), *Chislennost' i sostav naseleniya SSSR: po dannym vsesoyuznoy perepisi naseleniya 1979 goda,* Moskva, Finansy i Statistika.

Uzbek popular front (1989), *Central Asia and Caucasus Chronicle,* 8 (6), pp. 4–7.

Vakar, N. (1956), *Belorussia: the making of a nation,* Cambridge, Mass., Harvard University Press.

Weiner, M. (1978), *Sons of the soil: migration and ethnic conflict in India,* Princeton, N.J., Princeton University Press.

Williams, C., Smith, A. (1983), 'The national construction of social space', *Progress in human geography,* 7 (4), pp. 502–18.

Zaslavsky, V., Luryi, Y. (1979), 'The passport system in the USSR and changes in Soviet society', *Soviet Union,* 6 (2), pp. 137–53.

Zinchenko, I. (1984), Natsional'nyy sostav naseleniya SSSR, in *Vsesoyuznaya perepis' naseleniya 1979 goda,* Isupov, A. and Shvartser, N. (eds), Moskva, Finansy i Statistika, pp. 150–62.

4

Restructuring the Soviet city

Denis J.B. Shaw

For too long we disregarded the priority of the social sphere, putting off attempts to solve urban problems until 'better times'. What we did not understand was the fact that such times would never come as long as we neglected the needs of our cities (Kogan, 1990, p. 196).

Introduction

Complaints about the problems of living in Soviet cities are by no means new. But it is undoubtedly true that the analysis of past and present errors has never been more searching than it is now. The description by one group of scholars of the afflictions of the big Soviet city is fairly typical (*Krupnyy gorod,* 1988, p. 6). Among the social problems they describe are all kinds of deficits in the provision of goods and services, shortages in housing, social inequalities in various forms, anti-social behaviour, a growing amount of time spent on the journey to work and other difficulties. The worsening ecological situation, they say, is having a negative effect on the health of the population, the long-term consequences of which are by no means fully understood. And among the serious economic problems of big cities they name failures to invest adequately in productive capacity and infrastructure, waste of resources and ill-founded industrial location policies. What is ironic is that all these many problems have arisen in the context of an urban planning system specifically designed to make their solution easier. Yet these critics argue that the system is in fact incapable of providing solutions or of avoiding the negative impact on cities of the workings of the centrally planned economy. Indeed, they assert that urban problems have now become so severe that they are undermining the contribution which big cities could be expected to make to national economic performance.

What makes the existence of severe urban problems so significant for Soviet society is the central role cities play in the life of the nation. Nowadays two-thirds of the Soviet population lives in towns and cities (Table 4.1 and Map 4.1). Two out of every five Soviet citizens and three out of every five urban dwellers live in big cities with populations in excess of 100,000 (Tables 4.1 and 4.2 and Figure 4.1). And the fifty-six biggest cities with over half a million people account for a third of the urban and more than a fifth of the total population of the Soviet Union. It is in the towns and cities, and especially in the bigger cities, that the country's major industries, scientific and cultural potential, and key political and administrative personnel are located. A crisis in the cities is, therefore, a crisis in Soviet society as a whole.

Since Mikhail Gorbachev came to power in 1985, the problems of Soviet cities have if anything grown worse. The attempt to reform the Soviet economy, and with it many facets of the social and political system, has abetted many new difficulties without necessarily resolving the old ones. Nevertheless, the basic problems of the present-day Soviet city have their roots deep in history and it is thus necessary to examine that history before their character can be fully understood.

The Soviet experience of urbanism

Like human society as a whole, cities are a product of history. And since history has varied across the globe, so cities differ one from another. Attempts to generalize about cities are often fruitful in drawing our attention to what they have in common but they are also dangerous if they lead us to overlook the differences or to imagine that those differences do not matter. In the case of the modern Soviet city, the evolutionary process has been so distinctive that it is particularly important to bear the historical dimension in mind.

At the time of the Russian Revolution in 1917, only about 17 per cent of the Russian population lived in towns and cities. Russian society was quite underdeveloped in an economic sense, especially by comparison with such Western states as Germany, Great Britain and the United States. Russia's cities were mainly small and had many rural characteristics: provincial centres were often likened to overgrown villages by foreign visitors. Even the big cities like Moscow and St Petersburg (the modern Leningrad) were a mixture of modern and traditional in a way reminiscent of many Third World cities today (Hamm, 1976; Bater, 1976). A good deal of the industry was located in the countryside. Nevertheless, industrialization had made important progress since the 1860s and was changing the character of Russian society and its cities in important ways. Had the process been allowed to continue unhindered, it may be that its

Table 4.1 *Urban, rural and total population of the USSR*

Year	Total (millions)	Urban (millions)	Rural (millions)	Urban (per cent)	Rural (per cent)
1959	208.8	100.0	108.8	47.9	51.2
1970	241.7	136.0	105.7	56.3	43.7
1979	262.4	163.6	98.8	62.3	37.7
1989	286.7	188.8	97.9	65.9	34.1
1990	288.6	190.6	98.0	66.0	34.0

Source: Narodnoye khozyaystvo SSSR v 1989g (1990) Moscow, Finansy i statistika, p. 17.

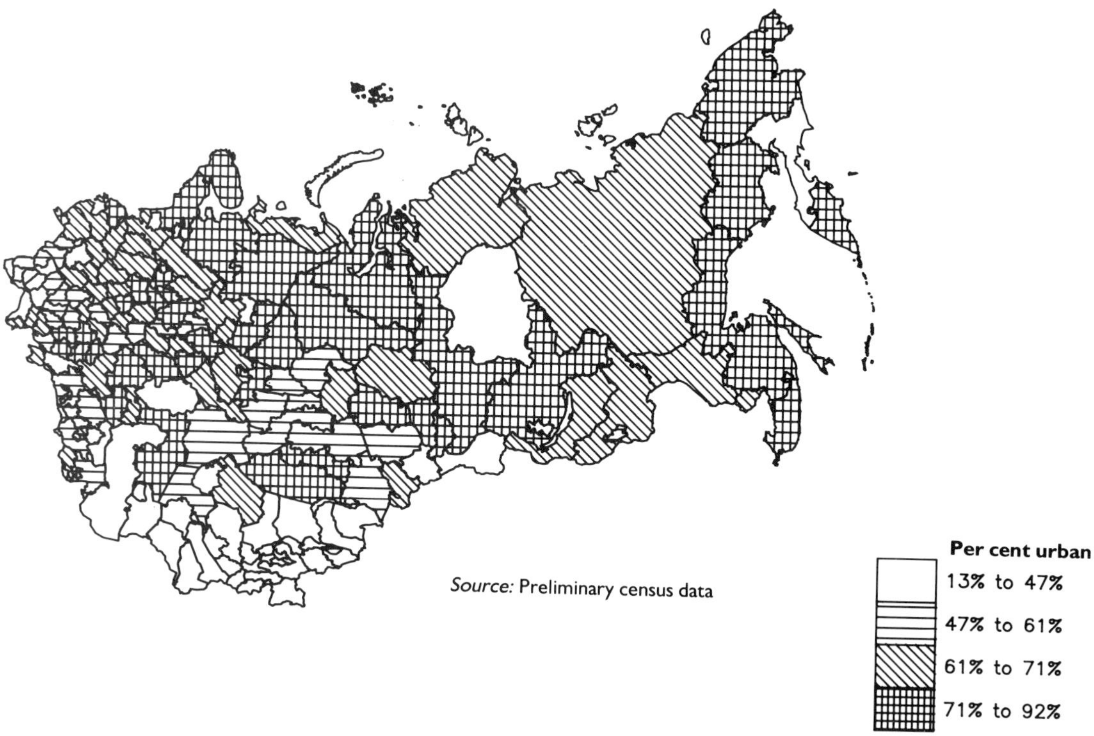

Map 4.1 *USSR levels of urbanization, 1989*

Table 4.2 *Distribution of urban settlements by population size*

	No. of urban settlements by year				Population (millions) by year			
	1959	1970	1979	1989	1959	1970	1979	1989
Total	4619	5505	5914	6216	100	136	163	188
Size categories:								
< 3,000	843	1118	1215	1301	1.6	2.1	2.1	2.3
3–4.9 thou.	904	1028	1081	961	3.6	4.1	4.3	3.8
5–9.9 thou.	1296	1430	1483	1515	9.2	10.1	10.6	10.9
10–19.9 thou.	798	919	987	1145	11.2	12.7	13.6	16.0
20–49.9 thou.	474	600	653	720	14.8	18.5	19.9	22.1
50–99.9 thou.	156	189	225	278	11.0	13.0	15.1	19.0
100–249.9 thou.	91	141	163	162	13.9	21.6	24.9	24.7
250–499.9 thou.	32	47	64	78	10.5	16.6	22.7	26.5
500–999.9 thou.	22	24	25	33	15.1	18.2	17.6	21.4
> 1 million	3	9	18	23	9.1	19.1	31.7	41.0

Source: Narodnoye Khozyaystvo SSSR v 1989g (1990) Finansy i Statistika, Moscow, p. 17.

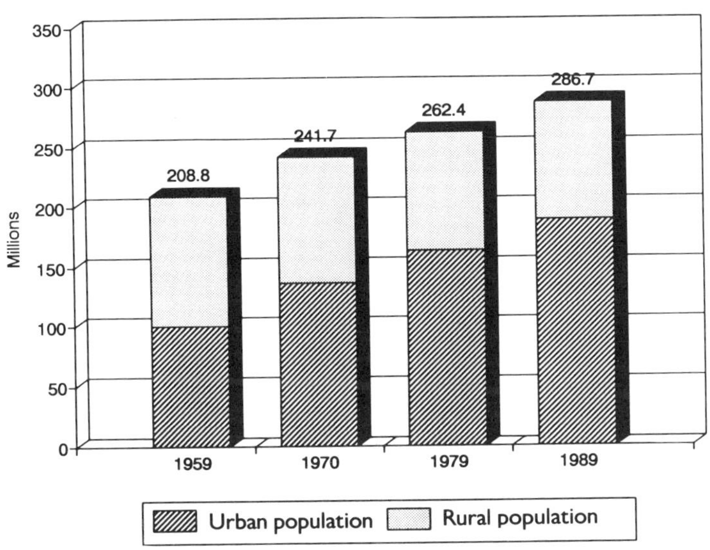

Figure 4.1 *Urban and rural population of the USSR*

big industrial cities would have evolved along similar lines to major capitalist cities in the West. Even before 1917, however, there were some features of Russian society that made it rather different from the West: features which may well have endured under almost any social system. One was the important role the state traditionally played in the economy, no doubt a product of Russia's relative backwardness. Another was the highly centralized character of the state itself and the restricted autonomy allowed to most social institutions, including urban government (Shaw, 1987).

It would be difficult to exaggerate the sweeping social changes that occurred in Russia as a result of the 1917 Revolution. The new Bolshevik or Communist regime was committed to the abandonment of the aristocratic and capitalist policies of its predecessor and to building a new socialist order. At first there was considerable uncertainty about how this was to be done, but by the late 1920s the outlines of the new approach began to appear. Under Joseph Stalin, by then the Party leader, some of the old Tsarist traditions such as political and economic centralization began to reassert themselves, but in a much exaggerated form. Stalin determined to pursue a policy of industrialization at maximum possible speed, relying as far as he was able on the Soviet Union's own resources through central control and planning of all economic sectors. The market was now virtually eliminated and all important economic decisions were made at the centre and passed down through a hierarchy of planning bodies to local enterprises and other organizations. The result of the policy of forced industrialization was spectacular economic growth that especially benefited heavy industry and the military sector. There were, however, grave costs to be borne. The consumer, welfare and agricultural sectors all suffered to a greater or lesser extent. The consequences for the living standards, freedom and even the lives of many citizens were also dire.

Stalin's policies had a dramatic impact on the cities. His determination to industrialize at all possible speed meant it was necessary to make good use of the economic potential inherited from the past. The already-existing industrial cities and regions therefore grew rapidly. But the policy of economic self-sufficiency also meant that new resource centres had to be developed, often in remote and underpopulated regions, and new communications built to interconnect them with existing centres. Many new towns and cities therefore appeared on virgin sites, many villages grew into cities and new centres appeared in strategic locations. It is important, however, not to exaggerate the degree of control that Stalin and his associates exercised over the numerous social changes. The drastic policy towards agriculture, replacing the traditional peasant farms with huge collectives, fomented mass migrations towards the cities often against the wishes of the authorities. Efforts to regulate migration in the case of the biggest cities were only partially successful. The authorities also attempted to regulate the development of cities by augmenting and systematizing the city-planning machinery (Parkins, 1953). This again was only partially successful in controlling the chaos of urban life. The task was made even more difficult by the onset of war in 1941. Because of the limited capabilities of the leaders in Moscow relative to the hugeness of the task in hand, and the restricted abilities of the local authorities (educated personnel were still rare in this newly industrializing society), great reliance was placed on the industrial branch ministries to solve their own problems at local level. Industrial ministries, for example, typically built housing, ran services and even constructed entire cities for their own workers. The problem with this approach was the tendency for each ministry to act independently of all others, which hardly ensured good local coordination.

Urbanization under Stalin, therefore, had a number of special characteristics which were to be of considerable significance for the future. First, it was extremely rapid, which inevitably produced social problems such as severe housing shortages. Second, it was not particularly well coordinated or planned. Third, it was deliberately unbalanced, with heavy industry being favoured to the detriment of consumer and other sectors. Fourth, there was little scope for local authorities (the so-called Soviets) or the local community to influence events. Finally, the sectoral approach to urban development was favoured over the spatial, which meant that individual economic ministries had great scope for organizing their own services and facilities locally, but at the cost of great unevenness in economic or social development across the city or the region.

All these factors acting together would have made the Stalinist city distinctive. But there is one other feature of the period that also deserves a mention. The determination to build a new society on a socialist basis inclined the Communists to accord considerable significance to symbolism. Symbols of the past, such as Tsarist memorials, churches, monasteries and even street and city names were viewed with suspicion and often obliterated. (Moscow, for example, is said to have lost 294 out of 586 churches, mainly in the 1930s.) In their place came new socialist symbols: statues of Lenin, Stalin and other leaders, grandiloquent war memorials, red flags, Party slogans, exhibitions of achievements, streets named after Party heroes and stirring revolutionary events, new socialist forms of architecture. The movement reached its apogee towards the end of the Stalin period with an accent on monumental architecture in a neoclassical style. New complexes were formed around symbolic spaces such as squares and boulevards, providing the perfect setting for military and political display. The irony is that in his concern to create a new social consciousness Stalin revived some of the pomp and ceremony (military splendour, nationalistic memorials) of the old regime. The result was the creation of a city that was distinctively (and perhaps disturbingly) new, and yet with many echoes of the past.

The period between Stalin's death in 1953 and the advent of Gorbachev in 1985 may be said to be an era in which the attempt was made to build the socialist city in a rather more generous and humane way than that adopted by Stalin and yet very much within the framework he had created. Thus a greater emphasis was placed on the provision of housing, on consumer and welfare questions, and on trying to develop an efficient system of city planning. But the fundamental priority of industrial and military power was not seriously challenged; the overall centralization of the state was upheld, much in the way he had devised. Cities, industries and housing complexes proliferated, but the accent was on quantity rather than quality. Life for the average city dweller became easier but living standards, which were rising quite quickly throughout the 1950s and 1960s, rose ever more slowly as time went on. By the 1980s it had become clear that the Soviet city dweller was still living a deprived life by Western standards and that the gap in living standards was unlikely to narrow in the near future.

Varities of urbanism

By the 1980s the Soviet experience of urbanization had produced a certain uniformity in towns and cities throughout the Soviet Union. The accent on socialist symbolism had created similar monumental spaces, similar memorials to Lenin and other worthies, similar slogans, street names and signs. Some of the distinctiveness of the older cities had been lost through demolition, wartime destruction and mass-housing projects. Uniformity of appearance is part of the unfortunate heritage of the industrial society, but in the Soviet case its impact was intensified by the speed of the industrialization and urbanization processes. Factory-made housing, mass-produced street furniture and a cleaving to uniform standards in many kinds of provision made the task of creating a varied and pleasing urban landscape well nigh impossible in many cases. Unfortunately, even where architects and history combined to create a happy effect, maintenance costs often conspired against the permanence of the achievement.

What Soviet urbanization failed to produce was any kind of uniformity in standards of living or social opportunities. Although much of the old social structure was swept away after 1917, the process of industrialization under Soviet conditions gradually created new inequalities, both social and spatial. Spatially, for example, there remained a sharp distinction between urban and rural lifestyles, arguably exacerbated by the pro-urban and anti-rural policies pursued for much of the Soviet era. Also important are the differences between cities. A number of attempts have been made to categorize these, although the lack of statistical data is a considerable problem. For the 1960s, Chauncey Harris used unpublished labour-force data to classify larger cities (Harris, 1970). He was able to show that while most of his cities were essentially industrial, a large minority had broader economic bases; these he described as diversified administrative centres. A similar exercise was undertaken by Olga Medvedkov (Medvedkov, 1980).

In a more recent study, Medvedkov used a broader range of data to classify big cities according both to their industrial profiles (using employment statistics) and to certain social development characteristics in the period since 1970 (Medvedkov, 1988 and 1990). She was thus able to classify cities into four groups. Of the 221 cities in her sample (i.e. those cities with over 100,000 people in 1970), 28 fell into her category of cities with restricted growth and significant socio-cultural development. This favoured group includes many bigger cities and republic capitals whose population expansion has been curbed by official restrictions but whose political and economic importance has encouraged the development of services and supporting infrastructure. Medvedkov believes that these cities benefit primarily from their centrality within communications networks. More dynamic are the 42 cities with significant growth and socio-cultural development. These cities show a spatial bias in their distribution towards the western and south-western borders of the Soviet Union, areas whose frontier location and dense agricultural populations (and thus adequate labour supply) have encouraged the development of many modern industries and supporting urban services. (Map 4.2 shows regional variations in the rate of urban population growth during the 1980s.)

Less well-placed are Medvedkov's remaining two categories of city. Fifty-nine cities are categorized as having significant growth but restricted socio-cultural development. This is a widely scattered group but Medvedkov argues that many of them are marginal to the major economic and political centres of the country, with a distinct spatial bias towards the east and the south. Typical of these will be the 'company towns' of the resource-frontier regions, cities developed and run by industrial ministries that are all too often more concerned with meeting industrial output targets than in providing comforts for their workers. Consequently, such cities are often characterized by inadequate social facilities, a narrow range of career opportunities, a transient and unsettled population and many social problems (Shelley, 1984).

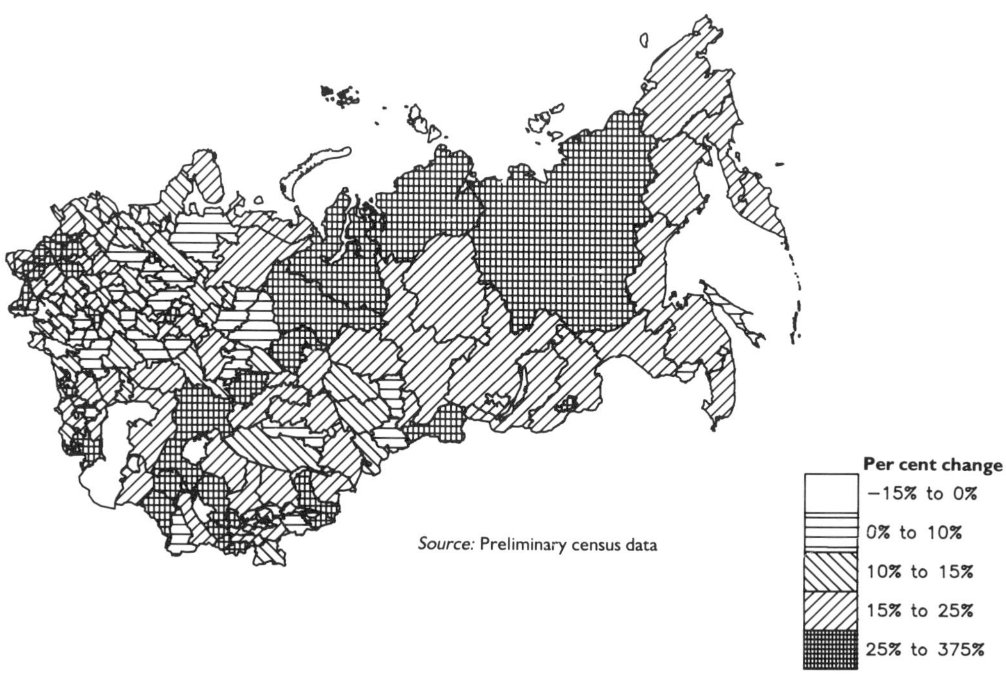

Map 4.2 *USSR urban population growth, 1979–89*

Finally there is the large group of 92 cities (42 per cent of the sample) with restricted growth and socio-cultural development. This scattered group of patently stagnant cities is regarded by Medvedkov as a casualty of the Soviet road to industrialization. These cities have outmoded economic structures and are often hampered from further development by chaotic land-use patterns and competition for space from neighbouring settlements. Clusters of these cities are found from the map to be situated in some of the most dynamic regions of the early five-year plans: the Donets-Dnepr region, the Urals, the Kuzbass and older parts of the Central Industrial Region. Analysis of data from the 1979-89 intercensal period shows a similar pattern, some cities actually losing population during this period (Cole, 1990). The reduced national dependence on coal production and steel making is an important causal factor.

The larger cities of the Soviet Union, those with over 100,000 population, are the ones most frequently discussed by scholars since they are the cities for which at least some statistical data are available. Smaller towns and cities are less well-known; this is especially unfortunate since their problems are frequently of a different order from those of the bigger city. Soviet urbanists talk of a 'small-town problem' which particularly afflicts those regions of the country far from more urbanized territories and where all the towns are small (Khorev, 1972). These towns are often characterized by semi-rural conditions, including narrow and inadequate employment opportunities, poor rural-type housing, inadequate services and other problems. For years it has been official policy to found branch plants in such centres or to develop them as agricultural processing and servicing centres. But they have suffered from ministerial reluctance to undertake the expense involved, their underdeveloped infrastructures and the poverty and backwardness of agriculture in many regions. Smaller towns and cities are obviously better off if they are located close to an urbanized region or at least to a communication link. Then at least it may be possible to persuade a firm to open a branch plant or affiliate nearby, or for individuals to commute to a nearby city. But the length of the journey to work in the case of many small towns (a task usually accomplished in the Soviet Union by rail) is often such as seriously to undermine their quality of life.

Yet another variable to Soviet urban life is the ethnic dimension. Despite Stalin's determination to 'Sovietize' the city (in practice this often meant to Russify it) and to erase much of the traditional culture, he barely succeeded in bringing any kind of uniformity to the many republics, autonomous regions and districts. National feeling was repressed rather than eradicated; today it is reasserting itself more stridently than ever. This is especially true of those non-Slavic republics where contact with the Russians has been more tenuous or short-lived. Cities naturally reflect the cultural context in which they are situated. This is why urban life today varies from the Europeanized milieu and lifestyle of the Baltic republics (Estonia and Latvia reflect their long associations with the Scandinavian and Germanic worlds) to the ancient Christian cultures of the Transcaucasus, and to the part-Islamic and traditional cultures of Central Asia (in the latter case, the level of urbanization is low by Soviet standards). Even within the Russian Federation, where the Russian nationality is generally dominant, ethnic differences have their effect (witness the recent reassertion of national feeling among the Tatars along the middle Volga, including Kazan', or among the northern native peoples of Siberia). One of the most serious shortcomings of the traditional Soviet approach to urban development was its insensitivity towards, or at most selective acquiescence in, this cultural variability.

At least some of the problems of Soviet urban development might have been eased had there been a fundamental attempt to construct a regional policy framework to counteract some of the unevenness. In fact, despite a good deal of rhetoric, little was done in this direction, at least in terms of comprehensive policy (Pallot and Shaw, 1981, pp. 55–86). Much was left to the branch ministries to decide and it was too easily assumed that the move towards socialism would somehow solve any regional difficulties without a great deal of thought or effort. This assumption has attracted a considerable amount of criticism recently, particularly in view of the continuing and widespread evidence of regional differences in living standards (Trofimov and Khakhulina, 1987; Gachechiladze, 1988; Hamilton, 1989). This is one important factor behind the current rise of regional and national consciousness threatening the overall unity of the country.

Problems of 'extensive' urbanization

It has already been noted that Stalin bequeathed to his successors a centrally planned economy in which each economic sector or branch was managed by a government ministry. The centre would draw up plans for the economy as a whole and the plans would be disaggregated into commands which would be sent to every ministry. It would then be up to the ministry to draw up plans for each firm or enterprise under its jurisdiction. This type of economic system, which worked reasonably well in the early years, began to develop a number of characteristic faults as time went by and as the economy became more complicated. One serious fault was the problem of innovation. It began to prove remarkably difficult to create a system of incentives that would encourage ministries and their enterprises to pay sufficient attention to research and development. In theory it might have been possible to do the research and development at the centre and then simply to command the ministries and their enterprises to adopt the new ideas. But in a modern economy there is an intimate link between research and development on the one hand and the actual production process on the other. Thus the centre was to some degree reliant upon the ministries to do this work. In practice, the very pressure placed upon ministries and their enterprises to fulfil their plans meant they had little time or inclination to attend to innovation. It was accordingly given a low priority. In the mid-1970s the Soviet authorities declared their intention of switching their economy from what was known as an 'extensive' basis, which meant having to rely on commensurate increases in inputs for each extra unit of output (hence the growing demands being made on expensive northern and Siberian resources), to an 'intensive' basis, which placed much greater emphasis on making more efficient use of the given inputs through technical innovation. Yet the inbuilt conservatism of the centrally planned system militated against this desired change.

The problem of innovation in the centrally planned economy is part of the broader problem of sectoral or ministerial autonomy. As the economy grew ever bigger and more unwieldy, it became increasingly difficult for the central planners to keep tight control over the ministries and their enterprises. It became necessary to rely on the ministries to undertake many tasks for themselves. This was often sensible, since the ministries were closer to the actual production process and knew the problems of their sector better than the centre. But it carried with it the danger that they would behave in what the centre considered sub-optimal or irrational ways. One problem area, for example, was that of investment. Ministries tended to be considered the best agencies to decide on the wisdom or otherwise of detailed investments; often even big investments were made without too much detailed control by the centre. Yet this frequently meant that ministries would aim to build their factories and plants in what seemed to them to be the best locations, notably those where plans could be fulfilled most easily. Yet this was not necessarily what the national interest seemed to require. Similarly, the ministries were left with a good deal of discretion over such matters as providing housing and services for their own workers and over labour recruitment, but this too might not accord with spatial planning requirements. Finally, because of the unwieldy character of the economy, the flows of raw materials and finished products were often disrupted, especially those flows that took place between sectors. This tended to encourage each ministry to try to become as self-sufficient as possible, even if this meant wasteful duplication and inefficient use of resources. All these problems could be counteracted by means of orders sent by the centre or alternatively by the use of what are known as 'economic levers' (fines, monetary incentives). But in practice it was simply impossible to attain the necessary degree of control over the situation.

A corollary of the fact that power in the centrally planned economy belonged to the central authorities and the sectoral ministries was the lack of power in the hands of the regional and local authorities – republics, regions, local Soviets including city Soviets. The latter had little or no control over what happened in their territories and relatively little ability to raise their own finances to develop services and production facilities. Although from the 1950s increasing efforts were made to control ministerial activity through various forms of spatial planning (Pavlenko, 1984; Shaw, 1985), the lack of local power to enforce spatial plans was a serious defect.

The structure of the centrally planned economy helps to explain many of the urban problems that have developed in the post-Stalin period. The constraints placed upon ministries and their enterprises have caused them to be cost-conscious to some degree and to try to avoid locating in expensive places without good infrastructures and adequate supplies of labour. Hence the problem of the small towns, already mentioned, and the tendency to avoid having to undergo the expense of development in, say, the north, Siberia or Central Asia where costs are high for environmental or social reasons. Where ministries have had to locate in places like Siberia to develop needed resources, the tendency has been to avoid unnecessary expense by cutting down on service and housing provision, recruiting an often young and transient workforce for high wages instead (hence the social problems of 'company towns' mentioned above). The forces that cause ministries to seek to avoid peripheral locations are the same forces encouraging them towards the more attractive sites, particularly the big cities in the European Soviet Union. These often have a good infrastructure, a varied and skilled labour pool, and other advantages. The fact that the ministries already control factories, housing and other facilities in these cities is a clear advantage, since by locating new facilities there they can avoid some of the inter-sectoral coordination problems described above. Not all the big industrial cities of the European Soviet Union are attractive for investment, for there are numerous stagnant or declining ones. These cities are victims of technological change (such as the switch from coal to oil and gas), though it must be added that the inherent conservatism of the Soviet economy has in many cases prolonged their economic lives beyond what might have been expected under market conditions.

Ministerial interest has therefore put great pressure on some cities while leaving others on the sidelines. Within the bigger cities in particular many problems have resulted. Ministries have adopted an 'extensive' approach to economic expansion, building additional production facilities and expanding their labour forces even where city plans forbid it. This results from the failure to innovate: because of the inability to make required savings in labour and resource use, the only way ministries and enterprises can fulfil their ever more ambitious plans is to acquire more resources and employ more labour. Despite laws that aim to control migration into big cities, it has proved quite impossible to restrain it or to prevent long-distance commuting from other settlements. Urban agglomerations have grown in an uncoordinated manner as enterprises, denied access to central cities, build on their peripheries, putting pressure on green belts and nearby cities. Uncontrolled migration and economic expansion have caused acute shortages in urban housing, services and infrastructures; such shortages have been exacerbated as the Soviet Union's economic difficulties have reduced the availability of investment capital. Ministerial licence has violated land-use controls, undermined the laws on urban conservation and pollution and frustrated the need to coordinate ministerial and local Soviet investment in urban development. Not everything can be blamed on the ministries, however. Many local Soviets, eager to attract ministerial investment, have connived at the infringements of city plans. It is often a choice between losing some planning goals through ministerial activity and losing other, perhaps more important ones, because of financial constraints.

Problems with urban transportation, land use, construction, servicing and environmental conservation naturally impinge on economic performance in all kinds of ways, which is why they have kindled a growing concern since the 1970s. Among the most important hindrances to economic performance are the effects of urban problems upon the labour force. Excessive competition for labour in big cities obviously makes recruitment difficult and encourages high turnover. Difficulties with shopping and services tend to foment absenteeism; efficiency is impaired by lack of incentives, low morale and fatigue. The latter is related to such factors as excessive journeys to work, queueing at shops and for services, and domestic stress. Pollution and other environmental problems are detrimental to the health of the workforce.

Urban problems thus undermine the efficiency of the labour force. It may also be that incentives are eroded and negative attitudes stimulated by social inequalities, especially when these are fostered by Party or bureaucratic privilege, and by dishonesty rather than by merit. A good deal of scholarly work suggests the presence of spatial inequalities in the Soviet city (Barbash, 1983; French, 1984; Bater, 1986; Sidorov, 1991). Needless to say, such inequalities hardly accord with the spirit of socialism and, if carried to excess, as they may have been in the latter part of the Brezhnev period, they threaten the legitimacy of the system. In the latter part of the Brezhnev era there was a noticeable increase in petty crime and lawlessness, including corruption in high places. A long-standing problem of the Soviet system has been the need to reconcile the goal of social justice and equality with the provision of material incentives to achievement. Reliance upon material incentives, however, can in certain circumstances serve to stimulate crude materialism. It is probably significant that these years also witnessed a sharp rise in black-market activities.

One theoretical advantage of the centrally planned economy has been its ability to husband resources and direct them at priority targets. This was the central achievement of the Stalinist years resulting in extraordinary rates of economic growth. It is an approach that has been resorted to again and again. As far as urban development is concerned, its most notable manifestation has been in the housing drive from the mid-1950s onwards, when massive resources were directed towards the solution of the housing shortage using centrally directed construction agencies and mass-production techniques. The consequence was gigantic strides forward in terms of quantitative achievements, though quality and sensitivity to local conditions inevitably suffered. It is an approach now regularly denounced by commentators as a typical example of the 'extensive' approach to urbanization (Platonov, 1989; Kolpakova, 1989; Kogan, 1989; Nekrasov, 1989). The result, they claim, has been the creation of monotonous, alienating townscapes and, in the adherence to standardized procedures, a failure to appreciate the varied needs of different segments of the population. The need for a 'community' approach has therefore been advocated. It could be argued that it is the failure to allow for local spontaneity and creativity that is one of the central shortcomings of the centrally planned approach. This failure has been a central target of Gorbachev's reforms.

Perestroyka and the Soviet city

In the years immediately following Gorbachev's rise to power in 1985, his policy of perestroyka appeared to have a number of specific objectives. The first was to reform the economy and thus to speed up the process of economic growth. The second was to democratize the system and secure public support for political and social change. The third was to maintain the Soviet Union's international prestige and move it forward towards a genuinely socialist society. The various reforms had many implications for the Soviet city, as we shall see. The problem is that as time has gone on and various difficulties have accumulated many of the original goals have been widely challenged and the legitimacy of the system has been brought into question. The very direction of perestroyka, and with it that of the Soviet city, is therefore a matter of great uncertainty. All that can be done in these circumstances is to discuss some of the most important reforms and suggest what their implications might be for the future of the Soviet city.

One of the central problems of the Soviet city has been the power of the industrial ministries and their ability to interfere with urban development in many ways. Their lavish use of resources reflects an insensitivity to costs because, although ministries have traditionally avoided 'unnecessary' expenditure (such as investing in peripheral locations), the centrally planned (administered) prices that guide them bear only a remote relationship to real costs (Buck and Cole, 1987, pp. 14–15). Thus decisions that appear economically rational to ministries or enterprises may be far from rational from the perspective of the economy as a whole. Part of the aim of economic reform has therefore been to bring prices into line with costs and to cause enterprises to be more sensitive to profit and loss through market-type stimuli. As far as the cities are concerned, for example, it has been argued that enterprises would become more sensitive to the value of urban resources if they were charged a proper rate for their use of land, water, electricity, and roads, fined for causing pollution and taxed on their use of scarce labour resources. Thus 'economic levers' might be used to cause enterprises to behave in the desired manner. The situation would then begin to move towards that found under capitalism where firms desiring to locate, say, in popular locations must pay for the privilege. This is a very different state of affairs from that prevailing in the Soviet Union until recently, since most resources, being publicly owned, have been free to the user. Traditionally, resources were allocated by plan or command but this is not particularly effective. Several cities are currently beginning to experiment with 'economic levers' (Karimov, 1989; Nekrasov, 1989; Chernik, 1990).

However, the use of 'economic levers' is by no means straightforward. For example, there is the question of the level at which to set rents, charges or fines. If they are too low, they will be ineffective; if too high, they may interfere with profits. With state-owned enterprises in effect paying the state for resources, there is the danger that fines or charges will be set at a 'manageable' level, so as not to impede what the enterprise considers expedient behaviour. Much therefore depends on how far the economy is marketized and how far enterprises are privatized. Under full marketization, however, there may still be a problem: market-type prices may stimulate firms to act 'rationally' as far as their own material interests are concerned, but this may not always be in the interests of good city planning.

Even under a full market economy, therefore, there will still be a need for effective city planning. Thus there arises the necessity of making it more efficient and increasing the powers of the local Soviets. A start to the former process was made in September 1987 with a decree on the further development of architecture and city planning (O dal'neyshem razvitii, 1987). Among a number of measures, this decree revamped the central planning agencies, and attempted to move away to some extent from standardization. Complaints are still being voiced, however, that the Soviet Union lacks proper city-planning legislation, meaning that city and other plans can be infringed with impunity (Kolpakova, 1989). The Union of Architects is said to be preparing legislative proposals. It is also encouraging the setting up of small architectural firms and building organizations, partly on a cooperative basis, to encourage individualized projects (Platonov, 1989; Kolpakova, 1989). It is too early to say how well this will work, but it is potentially important as one way of reducing the monotony and predictability of urban landscapes. In the meantime, the use of political slogans and posters on buildings and streets began to diminish soon after Gorbachev came to power. The deStalinization of the city was thus proclaimed in a very visible way. Communist memorials have also been removed in some places: the idea of using the townscape to inculcate communist values is therefore being seriously challenged, though not without the opposition of the communists themselves.

As regards the local Soviets, some progress has been made towards increasing their powers and ability to speak for local opinion by democratizing the election process, beginning in the spring of 1990. In theory at least, the Soviets should no longer be subject to manipulation by local Party bosses (the Party has now renounced its monopoly on power although it may retain its influence in some regions). The election of radical city governments in Moscow, Leningrad and other cities has provided the central authorities with numerous headaches and stimulated demands for further reform. Much remains to be decided about the power of the local Soviets. On the one hand, for example, the legal relationship between the city Soviets and those Soviets administering city boroughs has yet to be determined. Both are currently claiming the power to control local resources. On the other hand, the city's relation to its superior Soviets (the region and the republic) is also undecided, with numerous republics claiming autonomy or even independence from the Soviet Union's government. The question of city finances is another issue. The use of 'economic levers' has already been discussed, and this is linked to the concept of making cities and regions financially autonomous and economically fully accountable. Yet it is unclear at present how this would work: in the absence of some redistributive mechanism, its effect must presumably be to widen existing economic inequalities between cities (Artobolevsky, 1990).

The overall effect of the Gorbachev reforms has therefore been to move the cities some way in the direction of the market-type pluralism typical of Western societies. This tendency has been strengthened by much recent legislation affecting the life of the cities. Starting with the laws on the state enterprise (1987) and on cooperatives (1988), the setting up of cooperative and even privately-owned workshops, retail outlets and other services is now permitted. Opposition by bureaucrats and also by the public, who resent profiteering in conditions of economic scarcity and inflation, has proved formidable but at the time of writing by no means decisive. More recently the privatization of state-owned industrial enterprises has also been permitted, perhaps leading to investment by foreign firms. The establishment of joint ventures with foreign organizations has been allowed since 1987, and this has had some impact on the service sphere. If special economic zones are set up on the Chinese model in such locations as Vyborg, Leningrad and Nakhodka in the Far East, then these cities could experience much greater foreign involvement. Yet the gradual opening up of the Soviet economy to the outside world has been greeted by that world with an understandable caution. Economic difficulties and political uncertainties tend to discourage foreign investors.

Part of the political uncertainty derives from disagreements between the Soviet Union's constituent republics and the all-Union authorities, for example over economic legislation. Thus Soviet Union laws liberalizing the state control of land and property have been challenged in certain republics by more radical legislation. Republics are arguing with the centre over the control of resources. Privatization of property and housing is very much on the agenda, but exactly what the implications for the cities will be cannot be determined until the broader constitutional issues have been resolved.

Much of the current legislation is obviously designed to encourage enterprise and initiative, to shift some of the burden of maintaining the urban infrastructure from the shoulders of the state, and to give people a greater stake in their environment. What is not clear is whether such measures will make the city easier to plan, and whether they will succeed in creating a humane urban society. There is the obvious danger that marketization will increase social inequality in new and drastic ways, especially if it is accompanied by mass unemployment. Welfare arrangements under any new market system have yet to be agreed; the fear of impoverishment is widespread. Crime is already on the increase and there are many rumours of the advance of the organized variety. The instruments for maintaining social peace are being severely tested in the new conditions. Moreover, in the absence of a proper regional planning framework, there is a real danger that many cities and regions will suffer economic eclipse. Clearly, then, the move towards a market economy cannot be expected to solve urban problems overnight. Indeed, it may fail to solve any problems at all unless it is accompanied by many other measures, which is why the whole future of the reform process appears to be such an uncertain one.

Conclusions

The government of Mikhail Gorbachev has been attempting to grapple with the grave problems facing the contemporary Soviet city, problems which have been threatening not only the Soviet Union's position in the world but perhaps even the very foundations of the Soviet state. It has been argued above that the Soviet city's problems have their roots in history, in the way in which the Soviet Union attempted to industrialize and modernize itself in the first half of the twentieth century. The product of that history has been a distinctive set of urban problems, as well as other problems which the Soviet city shares with industrial cities anywhere. In conclusion it is appropriate to ponder how the Soviet city compares with other cities, for in this might lie a clue about how far present policies are likely to succeed.

In a previous article the present author argued that the Soviet city should be conceptualized as a distinctive type of industrial city, a 'state socialist' city as against the capitalist city that typifies the West (Shaw, 1987). Essentially, the argument is that the revolutionary changes that occurred in the Soviet Union after 1917, culminating in the Stalinist centrally planned economy and a new social structure, were sufficiently far reaching as to require new ways of theorizing about the Soviet city rather than the mere adaptation of theories about the capitalist city. For the geographer, one important question concerns the extent to which the distinctive character of the Soviet city is reflected in its spatial structure. This is not an issue to be examined here, except to suggest that there is some evidence that the spatial structure is distinctive in a number of significant respects (see French and Hamilton, 1979).

With the onset of Perestroyka, however, it is evident that the concept of the 'state socialist city' requires re-examination. If, as the present chapter has argued, the effect of Gorbachev's reforms has been to move the Soviet city in the direction of Western-type pluralism, does this mean that the distinctive character of the Soviet city will be entirely lost and that it will become merely another variant of the common capitalist type? This is obviously a huge question that can hardly be dealt with in a concluding section, and in any case the fate of the reform process is still extremely uncertain. What can be said is that the heritage of more than seventy years of Soviet-style socialism will not easily be discarded, no matter how desirable that goal might appear. In the article alluded to above, the present author argued that Russia's pre-revolutionary development, including its urban experience, cannot be ignored in seeking to understand the character of its modern cities. This pre-revolutionary heritage is apparent not so much in the physical form of cities (the majority of which, with certain very important exceptions, are either entirely new or have been radically altered) as in the behaviour of their institutions and in certain social attitudes. The old Russia, seemingly entirely eradicated after 1917, continued to exert a secret influence the importance of which is only now becoming apparent as many old traditions and attitudes begin to reassert themselves. If all this is true of old Russia, then it will undoubtedly also be true of the period after 1917. That, after all, was the period of the Soviet Union's industrialization, during which most of the physical fabric of its cities was constructed and its present social structure and economic system were formed. No matter how radical the economic transformation now facing the Soviet Union, no matter how sweeping the subsequent social reforms, the experience of Stalinist and post-Stalinist industrialization and urbanization will continue to mould the character of its cities in innumerable ways and for a long time ahead. To that extent the Soviet (or Russian) city will still be disinctive.

All this does not mean that Soviet cities may not experience radical transformation in the years ahead, though at present it is difficult to foresee what form that transformation may take. The path may be firmly set in the direction of Western-type pluralism or it may deviate to one side. The Soviet Union may be fully integrated into the world economy, or it may continue to retain much independence. The future of the city obviously depends on that of the Soviet Union as a whole. But whatever policy choices are made, whatever events intervene, the problems inherited from the past will hardly be solved quickly. As time goes by, perestroyka looks more and more like a formidable mountain: the slopes are steep and the summit is hidden by dark clouds.

Suggested reading

French, R.A. and Hamilton, F.E.I. (eds) (1979) *The Socialist City*, Wiley, New York.

Harris, Chauncy D. (1970). *Cities of the Soviet Union*, Rand McNally, Chicago.

Medvedkov, O.L. (1990). *Soviet Urbanization*, Routledge, London.

Pallot, J. and Shaw, D.J.B. (1981). *Planning in the Soviet Union*, Croom Helm, London.

References

Artobolevsky, S. (1990). 'The Capability of the State to Solve Regional Problems', *Geoforum*, 21 (4), pp. 449–54.

Barbash, N.B. (1983). 'Physical Development of Infants as an Indication of the Condition of the Urban Environment', *Soviet Geography*, 24, pp. 204–13.

Bater, J. H. (1976). *St Petersburg: Industrialization and Change*, Edward Arnold, London.

Bater, J. H. (1986). 'Some Recent Perspectives on the Soviet City', *Urban Geography*, 7, pp. 93–102.

Buck, T. and Cole, J.P. (1987). *Modern Soviet Economic Performance*, Blackwell, Oxford.

Chernik, D. (1990). 'Eksperiment: O printsipakh samoupravleniya i samofinansirovaniya g. Moskvy', *Gorodskoye khozyaystvo Moskvy*, 4, pp. 4–5.

Cole, J. P. (1990). 'Changes in the Populations of the Larger Cities of the USSR', *Soviet Geography*, 31, pp. 160–72.

French, R.A. (1984). 'Changing spatial patterns in Soviet cities: planning or pragmatism?', *Urban Geography*, 8, pp. 309–20.

French, R.A. and Hamilton, F.E.I. (1979). 'Is There a Socialist City?', in French, R.A. and Hamilton, F.E.I. (eds), *The Socialist City*, Wiley, New York, pp. 1–21.

Gachechiladze, R.G. (1988). 'Analysis of Regional Differences in Material Welfare of the Population: the Georgian SSR', *Soviet Geography*, 29, pp. 413–19.

Hamilton, E. (1989). 'The Social Contract and Well-Being in the USSR: a Comparison of Central Asia, Kazakhstan and the Donets-Dnepr Region', *Soviet Geography*, 30, pp. 109–29.

Hamm, M.F. (1976). 'The Breakdown of Urban Modernization: a Prelude to the Revolutions of 1917' in Hamm, M.F. (ed.), *The City in Russian History*, University Press of Kentucky, Lexington, pp. 182–200.

Harris, C.D. (1970). *Cities of the Soviet Union*, Rand McNally, Chicago.

Karimov, A. (1989). 'Sistemno podkhodit' k razvitiyu goroda', *Stroitel'stvo i arkhitektura Rossii*, 1, pp. 4–5.

Khorev, B.S. (1972). *Malyy gorod*, MGU, Moscow.

Kogan, L. (1989). 'Gorod i politika', *Stroitel'stvo i arkhitektura Rossii*, 1, pp. 8–9.

Kogan, L.B. (1990). *Byt' gorozhanami*, Mysl', Moscow.

Kolpakova, M. (1989). 'Genplan goroda: zakon?', *Arkhitektura i stroitel'stvo Rossii*, 9, pp. 8–9.

Krupnyy gorod: problemy i tendentsii razvitiya (1988). Nauka, Leningrad.

Medvedkov, O.L. (1980). 'Structural Characteristics of Cities as Multifunctional Centers', *Soviet Geography*, 21, pp. 263–83.

Medvedkov, O.L. (1988). 'Soviet Cities and their Industrial-Social Performance', *Urban Geography*, 9, pp. 487–517.

Medvedkov, O.L. (1990). *Soviet Urbanization*, Routledge, London.

Nekrasov, I. (1989). 'Gorod vne zakona', *Arkhitektura i stroitel'stvo Rossii*, 2, pp. 30–1.

'O dal'neyshem razvitii Sovetskoy arkhitektury i gradostroitel'stva' (1987). *Sobraniye postanovleniy pravitel'stva SSSR: otdel pervyy*, No. 45, st. 149, pp. 947–68.

Pallot, J. and Shaw, D.J.B. (1981). *Planning in the Soviet Union*, Croom Helm, London.

Parkins, M.F. (1953). *City Planning in Soviet Russia*, University of Chicago Press, Chicago.

Pavlenko, V.F. (1984). *Planirovaniye territorial'nogo razvitiya*, Ekonomika, Moscow.

Platonov, Yu.P. (1989). *Arkhitektura SSSR*, 4, pp. 2–3.

Shaw, Denis J.B. (1985). 'Spatial Dimensions in Soviet Central Planning', *Transactions of the Institute of British Geographers N.S.* 10, pp. 401–12.

Shaw, D.J.B. (1987). 'Some Influences on Spatial Structure in the State Socialist City: the Case of the USSR', in Holzner, L. and Knapp, J.M. (eds), *Soviet Geography Studies in Our Time*, University of Wisconsin, Milwaukee, pp. 201–27.

Shelley, L. (1984). 'Urbanization and Crime: the Soviet Experience' in Morton, H.W. and Stuart, R.C. (eds), *The Contemporary Soviet City*, Macmillan, London, pp. 113–26.

Sidorov, D. A. (1991). 'Perception of Variation in the Level of Living Among and Within Large Soviet Cities', Paper presented at the Annual Conference of the Institute of British Geographers, Sheffield, January.

Trofimov, V.A. and Khakhulina, I.A. (1987). 'The Standard of Living in Siberia Compared with other Regions of the RSFSR', *Soviet Sociology*, 25, pp. 17–34.

5

The countryside under Gorbachev

Judy Pallot

> I think that the farmer we are so afraid of will be a buttress of our perestroyka in the countryside. Stolypin was no fool. He understood who could be the foundation, so to speak, of the state (Adamovich, 20 February 1990).

Introduction

These words were spoken by Adamovich, a deputy to the Supreme Soviet, on 20 February 1990 (Summary of World Broadcasts, 20 February 1990). In many ways they encapsulate the attitude of radical reformers in the Soviet Union to the 'rural question': first, in the importance they place on individual, as opposed to collective enterprise in achieving a restructuring of the rural economy; and second in the recognition of the centrality of rural change to the outcome of reform in the Soviet Union in general. The mention of Stolypin refers to the reforms introduced at the beginning of the twentieth century by the then prime minister Petr Stolypin which sought to solve the dual problem of rural unrest and agricultural backwardness by substituting compact owner-occupied farms for the fragmented and communally owned land parcels of the peasant commune. Although it would be unwise to draw too strong a parallel between Stolypin's and Gorbachev's agrarian reforms, it is true that today's attempt to introduce individual farming into the Soviet countryside has the potential, as had Stolypin's, to effect nothing short of a full-scale revolution in Soviet rural society (Macey, 1990).

The preconditions for the rural revolution are already in place: first, there has been a general society-wide reappraisal of the value of rural life; secondly, legislation has been introduced to promote individual farming and other forms of private activity in the Soviet countryside; and thirdly, resources have been earmarked to upgrade rural standards of living.

The perceptual revolution

Collectivization and its aftermath

For seventy years the superiority of the collectivist principle in agriculture has been the *sine qua non* of the Soviet approach to rural development. Soviet policy-makers have shown their adherence to this ideal in the conscious choice of policies that have promoted joint forms of labour organization, responsibility and rural living, and have discriminated against all forms of individualism. The collectivization drive in the 1930s was the clearest example of these preferences; as is well known, it involved the surrender by formerly independent peasant producers of their land and means of production to large-scale collective farms (*kolkhozi*) and the pooling of their labour for work on the newly socialized land. The goal of each collective farm was to meet procurement quotas set by the state-planning organs, and the overriding goal of the agricultural sector was to provide the resources for high-speed industrialization. Effectively, collectivization eliminated the independent peasant farm from the Soviet countryside, although pockets of limited individual activity remained in the form of the 'private plot', allowed collective farm members for their personal cultivation.

In the decades following collectivization the commitment to the collectivist principle in agriculture was maintained. Reforms in agricultural organization introduced by the post-Stalin leaders tended to push agriculture further down the same road; collective farms were merged into ever-larger units, many were converted into state farms (*sovkhozi*) that were even more closely integrated into the state-planning apparatus, and, from the 1960s, joint enterprise between farms, or between farms and industry, was encouraged through the formation of agro-industrial corporations and inter-farm cooperatives (Table 5.1). Meanwhile, whenever overall levels of production allowed, campaigns were launched against the private plot as the last vestige of individual enterprise in the countryside (Wadekin, 1973). Inevitably, the changes in rural society brought about by the collectivization of agriculture were far-reaching. Former hierarchies of power were replaced by new hierarchies which placed party functionaries and officials in the state-planning organs in commanding positions. On the farms themselves the chairperson, specialists such as accountants and veterinary surgeons, and brigade leaders wielded most power. Peasants, now recast as *kolkhozniki* or collective farm workers, found

Table 5.1 *The area of agricultural enterprises in the USSR on 1 November 1989 (in millions of hectares)*

Type of agricultural enterprise	Total land area in the territory of the enterprise	Total agricultural land	Percentage share of total agricultural land
Collective farms of which	245.9	173.7	31.1
-land in collective use	241.4	169.5	(30.4)
-in the private use of collective farm members ('private plots')	4.0	3.8	(0.7)
-in the private use of other workers	0.5	0.45	(0.08)
State farms and other state enterprises	801.3	377.2	67.6
Interfarm enterprises and organizations	2.8	2.4	0.4
Individual citizens (excluding plots on collective farmland)	4.9	4.5	0.8
Total land in the use of agricultural enterprises	1055.0	557.9	100.0

Source: Narodnoye Khozyaystvo SSSR v 1989 g. (1990) Finansy i Statistika, Moscow, p. 432.

themselves working in labour brigades fulfilling plans into which they had had little effective input. Whereas in the pre-collectivization village peasants were differentiated from one another in terms of their land and livestock holdings, in the collective farm differentiation came to be associated more with party membership, formal position in the decision-making hierarchy and the type of brigade, for example mechanized or manual, to which a person belonged.

The rationale for these changes in the countryside rested largely on arguments about the economies of scale that could be achieved through pooling labour and non-labour resources. In addition, however, there was a strong element of pure anti-peasant sentiment in the Communist Party's decision to collectivize and in its subsequent adherence to this form of farm organization. The Communist Party's suspicion of the peasantry was rooted partly in the theoretical writings of Marx, Engels and Lenin, and partly in the historical experience of the peasants' 'unreliability' at various crucial points in the history of the revolution. Although there is strong evidence that Marx saw possibilities for a peasant-based transition to socialism in Russia, the message the Russian Bolsheviks chose to read from Marx was the historical inevitability of the peasantry's disappearance during the course of development and of their essential petty-bourgeois, counter-revolutionary nature. Thus, even though Lenin and the Bolsheviks were prepared to forge an alliance between the peasantry and the working class for the sake of the Revolution, the majority of Bolsheviks viewed the peasantry as the lesser partner in this alliance and a potential threat if left to its own devices (Kingston-Mann, 1985).

During the period of War Communism immediately after the revolution and later during the comparatively liberal period of the new economic policy in the 1920s, the relationship between the peasants and the first workers' state remained uneasy; the failure to achieve the level of agricultural production deemed necessary for the economic development and defence of the state reinforced mistrust of the peasantry in the party. With the majority of Bolsheviks holding, at best, ambivalent attitudes towards the peasantry, a policy that simultaneously promised to raise agricultural output and to eliminate the most extreme petty-bourgeois elements of the peasantry, while transforming the remainder into agricultural workers analogous to the industrial proletariat, fell on fertile ground. This was the promise held out by collectivization. During the course of 1939 to 1941 95.6 per cent of all peasant households in the Soviet Union joined collective farms. The immediate impact of the changes was the destruction of many of the structures and traditions of existing rural society, as well as the loss of life due to executions, deportations and famine. In collective farms peasants were firmly subordinated to the authority of the state and reduced to a position of 'social dependency'(Tikhonov, 1989, p.38). Although communist party textbooks have presented collectivization as 'the liquidation of the kulaks' (that is, the elimination of the most wealthy and, potentially, counter-revolutionary peasants), it is now recognized that the social consequences were much greater. In the words of Academician Tikhonov of the All-Union Lenin Agricultural Institute, de-kulakization not only subjugated, 'the most skilled, experienced and energetic farmers', but also served as the basis for a most tragic historical act – de-peasantization; that is, the elimination of the peasantry as a large social formation with its own place in production, its own social psychology and way of life' (Tikhonov,1989, p. 35).

In the decades following the collectivization drive, rural policy tended to reinforce the peasants' subordinate position in Soviet society either by overtly discriminating against peasants or by forcing urban and proletarian values upon them. An example of the former was the withholding from collective farm members of the internal passport necessary to any Soviet citizen if he or she was to change job and place of residence. Only in the middle 1970s did collective farm members become entitled by right to these documents and even then, in practice, their access to them was circumscribed. An example of the latter was the approach to the redevelopment of rural settlements. Until the 1950s little thought was apparently given to the type of settlement the collective and state farm worker was to live in, but from that time a programme was launched which tried to introduce the 'urban way of life' to rural areas. The programme was the brain child of Khrushchev who, fired by ideological conviction about the superiority of urban life, and determined to stem rising levels of rural out-migration, became convinced that villages and hamlets should be reconstructed in the form of agricultural towns. 'Rural settlements of the new type', which were to offer the same range of services as towns, were actively promoted by the communist party as being the most appropriate form of settlement on collective and state farms. The physical plans drawn up for such new rural settlements mimicked themes in contemporary town planning with blocks of flats the principal housing type chosen, a ceremonial square with its obligatory statue of Lenin at the centre and hotels, sports stadia, houses of culture and administrative buildings. In reality, for lack of resources, few of the Soviet Union's villages were reconstructed in this way, but the urbanization of the village remained the official goal of rural settlement planning for thirty years (Pallot, 1979, 1987 and 1988).

Rural policy reappraised

Before Gorbachev came to power in the 1980s the tenets underlying Soviet farm organization and rural development had begun to be questioned in a variety of quarters. Inevitably, political dissidents were the severest critics of the treatment of the peasants. Gregory Grossman's personal account of collectivization in *Vse Techet* graphically describes the horrors of the 1930s rural transformation; Roy Medvedev's *Let History Judge* attempts a quantification of lives lost and peasants deported in the 1930s as a direct result of the campaign. But from within official Soviet society criticisms were also voiced in the post-war Soviet Union. Most notably, there emerged a group of writers in the 1960s who came to be known as the *derevianshchiki*, who bemoaned the loss of the Russian village and with it the mainspring of society's 'spirituality and morality'. These writers have subsequently assumed a leading role in current debates about rural revival, de-collectivization and environmentalism although their case has been complicated by the undertones in it of anti-Semitism and its association with right-wing nationalism.

Outside literary and dissident circles criticisms of rural policies were concerned more with production problems than with living standards and the 'condition' of the peasantry. From the 1950s the allocation of resources to agriculture rose, so that by the middle 1970s the sector was commanding around one quarter of all new investment in the Soviet economy, and there were some reforms in the planning mechanism and in the methods of payment and labour organization on collective and state farms. However, none of the post-war leaders was prepared to question the validity of large-scale socialized farm units as the form of agricultural organization suited to the Soviet Union. Under Khrushchev and then Brezhnev, heavy investments were made in the mechanization and chemicalization of agriculture and in bringing new land into use through ploughing campaigns and land reclamation. These measures met with only partial success. Thus, while the volume of the harvest has expanded in post-war years, there have been sharp annual fluctuations in yield and imbalances in production that have kept pressure on the state to import from overseas (see Table 5.2 and also Chapter Nine). Furthermore, inadequate rural infrastructure and labour shortages have resulted in a growing uncertainty about the ability to bring the harvest in even if good: 1990 was a case in point. Where living conditions in the countryside were concerned, plans to build 'settlements of the new type' were of doubtful benefit. Their principal effect was to give the authorities an excuse to neglect small villages and hamlets and to concentrate services in selected larger villages.

When Gorbachev first came to power he acknowledged the relative failure of his predecessors' attempts to solve the agricultural problem but, like them, he was not prepared to call farm organization into question. Inevitably, however, the freedom of expression allowed under glasnost was soon used by critics to attack first the conduct of the collectivization drive and later the collectivist principle in general. Socialism was accused of removing responsibility for their labour from workers by separating them from the means of production and that 'in the village this separation, of people from the land, turned out to be more intense than in other branches of the economy' (*Arenda*, 1988, p. 3). As a result, it was argued, interest in agricultural labour had declined. This had led to falling work standards and mass rural out-migration: both, in turn, held to be major

Table 5.2 *Cereal harvests and imports, 1980–9*

Year	Gross output of grain: (in mmt)	(% growth)	Ceral imports (in mmt)
1980	189.1	–	27800
1981	158.2	–16.3	n.a.
1982	186.8	18.1	n.a.
1983	192.2	2.9	n.a.
1984	172.6	–10.2	n.a.
1985	191.7	11.1	44200
1986	210.1	9.6	26757
1987	211.4	0.6	30385
1988	195.0	–7.8	35042
1989	211.1	8.3	36000

Source: PlanEcon Report, 21 February 1990, p. 25.

contributors to rural economic and social problems. The consequences of rural out-migration for the fate of the Soviet Union's villages came in for particular scrutiny at this time. Taking up the theme of the rural prose writers, officials, journalists and academics now began to lament the disappearance of villages and village society in the heavily depopulated regions in the heartland of the Russian Republic (see Map 5.1). In the words of academician Yemel'yanov of the All-Union Lenin Agricultural Institute: 'Villages have been depopulated and their inhabitants are old. The centuries old tradition of peasant life has been destroyed. Habits of work and moral principles have been forgotten. It is essential that we do everything possible to save the village' (*Moskovskii Komsomolets*, 1989, p.2). Yemel'yanov's lament and the call for action to save the village can be said to represent a new orthodoxy with respect to rural development in the Soviet Union and they signal a rejection of the negative attitudes towards the peasantry that prevailed in the past.

For the first time in decades current Communist Party policy is attempting to bridge the gap between the narrow concern for agricultural production and broader problems associated with living in the countryside. This new approach is summed up in Gorbachev's speech to a plenum of the Communist Party convened to discuss the agricultural crisis in March 1989. In it he said that the principal lesson of history is 'the essential need for caring attitudes to the conditions of work and life of the rural labour force. The agrarian policy we are developing today must put as its first priority the genuine rebirth of the village, so that the peasant and agricultural work will be given a respected place in society. This is not only a most important socio-economic task but I would say our moral duty . . . to care about rural people is an essential part of our agrarian policy' (*Ekonomicheskaya gazeta*, 1989, p. 3). With these words the perceptual revolution towards the peasant can be said to be complete. However, as the next section will show, rural society is proving to be less receptive to change than optimistic reformers expected.

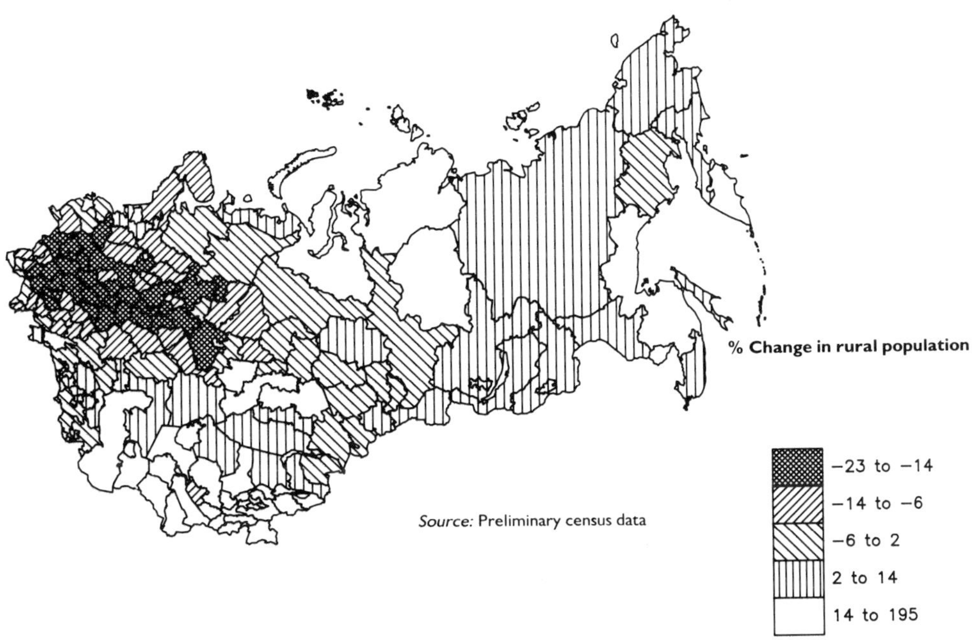

Map 5.1 *Rural depopulation in the USSR, 1979–89*

The privatization of the Soviet countryside

The clearest example of the changed attitudes towards rural life on the part of the Soviet government is in the measures it has passed to encourage the establishment of individual farming. The first moves in this direction were taken almost immediately after Gorbachev came to power in the form of a joint Council of Ministers and Communist Party decree designed to increase worker interest and material incentives in agriculture. Under the provisions of this decree it became possible for groups of farm workers, whether organized in small teams (*zvena*), or brigades, or constituting a farm sub-division to negotiate a contract with the parent collective to do a specific task or tasks on the farm. Such contracts provided for the payment of the product delivered to the collective farm at stable prices, fixed in advance for periods of five years, at levels up to 150 per cent of the normal rate, and for the wages paid to workers to be differentiated according to the effort they made. Contracting teams, brigades or sub-divisions for their part took on responsibility for covering costs under the system of cost-accountability, or *khozraschet*. The argument for these arrangements was that they would engender new attitudes on the part of both negotiating parties; farm managements would learn to respect 'freedom of choice', and 'to honour negotiated terms', while the collective contract workers would acquire 'technological discipline', take greater responsibility for their labour and, generally, benefit from a 'healthy moral climate' (*Slovar'-spravochnik*, 1990, p. 35). A version of this type of arrangement, actively promoted by Gorbachev himself at about the same time, was the family contract, or *semeinii podryad*, whereby relatives could join together to negotiate a work contract with their collective farm. The novelty of the family contract lay not in the terms negotiated, which were no different in scope from those possible with collective contracts, but in the promotion of the idea of a family work-unit with its connotations of the traditional peasant farm. Given the, by then, growing nostalgia about the peasant, it is not surprising that the family contract received much media attention and was believed to have an important role to play in the rural renaissance. According to two reporters writing in *Pravda*, for example, family contracts lead to 'the democratization of rural life and work; it is a means through which feelings of worthiness and that one is a master of one's own land and one's own fate, are increased' (*Arenda*, 1988, p. 78).

Whatever their symbolic novelty, family contracts did not sever the ties binding farmworkers to their collective and state farms. Under the contract system the groups of workers, however constituted, would still depend upon their parent farm for the means of production, fuel and seeds. Further, they would still be expected to spend some of their time working in the socialized sector and to dispose of the greater part of their product through their parent collective. Workers on the contract system had no real security of tenure over the land they worked and most contracts were, in any case, for relatively short periods of time – for ten to fifteen years. Essentially, therefore, the contract system represented a decentralization of decision-making within the framework of socialized production. It was another means for collective and state farms to meet administratively determined procurement quotas.

The desire to promote individual forms of enterprise within the framework of collective and state farms has also been reflected in measures designed to expand production in the Soviet Union's existing 'private sector'. A decree 'On additional measures for increasing the production of agricultural products from the subsidiary household plots of citizens' was introduced in 1981. This loosened some of the restrictions that previously had applied to the use of 'private plots'; further measures to encourage private plot production were introduced in 1987. Private plots are the small parcels of land, usually no greater than 0.5 hectares, to which farm members and certain other citizens are entitled for their own personal use. For the greater part of Soviet history these plots have been used for subsistence purposes (indeed, in the 1930s and 1940s they were essential for the collectivized peasant's survival since remuneration from the socialized sector was negligible), although a portion of the produce has always been marketed. Compared with the socialized fields, private plots are very productive; accounting for only 0.7 per cent of all agricultural land in the Soviet Union (see Table 5.1), they have consistently produced around 30 per cent of total output. This has been possible because of their specialization on high-value products, such as fruit and vegetables, and dairy products which the socialized sector has failed to produce in sufficient quantity and quality to satisfy consumer demand. The 1981 decree removed the formerly strict limit on the number of livestock that could be held on private plots, increased plot size and allowed for extra leasing of land from collectives in order to grow fodder crops (Medvedev, 1987, pp. 374–83). These are important changes but, as under the contract system, peasants remain dependent on their parent collectives and subject to strict rules about the marketing of their products.

A more fundamental move towards the development of private farming came in 1988 when measures were introduced allowing the individual or cooperative renting of land together with its buildings and appurtenances and certain movables, such as machinery, tools and farm vehicles. The intention at the time was that this change in the law would be used by existing farm workers. Its aim was to give some legal force to contracts negotiated by collective and state farms with their members, thereby increasing everybody's confidence in them. With the same end in view the upper limit on the length of time for which such contracts could be negotiated was set at fifty years. Under the terms of the typical rental contract, or *arendnyi podryad*, part of the product of the land had to be realized through the parent farm, at previously agreed prices, in addition to the monthly rent, payable in money or kind. Restrictions were placed on sub-letting and the use of permanent hired labour but, beyond these, people leasing land could expect to be left to make their own production and investment decisions free from interference by the parent farm. Cooperative renting was covered by a law on cooperatives which laid down, in addition to the above, rules governing the form, conduct and membership of cooperatives (*Slovar'-spravochnik*, 1990, pp. 246–54).

The Communist Party placed great hopes in its agrarian reforms; rent, contracts and strengthened 'private plots'seemed to offer a way of improving agricultural efficiency while not seriously challenging the collective- and state-farm system. Indeed, official texts in the 1980s were at pains to demonstrate the socialist nature of the changes. For example, at a July 1988 Central Committee plenum, it was declared that rent 'fully realizes the possibilities of socialist property' (*Slovar'-spravochnik*, 1990, p. 117), while at a 12 October meeting between the Central Committee and collective and state-farm managers Gorbachev declared that rent was the agent of 'socialist democratisation' in agriculture (*Arenda*, 1988, p.3). Collective farms have been portrayed as only benefiting from the changes since they are being returned to Leninist principles and their 'true cooperative nature' (*Slovar'- spravochnik*, 1990, p. 253). The independent farmers, meanwhile, are described as neither capitalist nor peasant but as 'a phenomenon of our times' whose appearance signifies the birth of 'the socialist farmer . . . an inventive organizer of production and a fully fledged proprietor' (*Arenda*, 1988, p. 100).

'Private' farming – the record to date

Early reports of functioning rented farms and work contracts seemed to confirm the belief that decentralization within the existing structure of collective and state farms could bring substantial improvements to the efficiency of agriculture. The experience, for example, of the Vyshne-Ol'shanski sovkhoz in Orel oblast (in the Central European region of the RSFSR) after it let out its land to a mixture of cooperatives, individual tenants and family teams is typical for reports in the mid- to late-1980s; namely, a 59 per cent increase in gross output by the end of the first year, a 50 per cent reduction in the size of the farm 'apparat', and a labour deficit turned into a 60 person surplus (*Arenda*, 1988, p.20). These improvements were attributed to the high level of labour discipline needed to cover costs (including wage costs) and to various 'psychological factors' which changed the farm worker from being 'an unthinking fulfiller of orders' to someone in whom 'the long dormant interest and initiative of the peasant farmer has been awakened' (*Arenda*, 1988, p. 55).

After the initial success stories, however, a more pessimistic tone began to dominate reports about the agrarian reform; evidently there were obstacles holding up the adoption of rental or other contracts or, where they had been formed, prevented their full benefits from being realized. The principal obstacle was resistance on the part of rural people themselves. At the highest level in rural society, among farm managers, local administrators and specialists, the resistance was relatively easily explained by the fear of losing power and of incurring extra work and high costs. Where contracts had been negotiated, constant interference by farm officials has been a problem. As one member of a collective farm in Moscow oblast reported, 'teams and brigades, all the same, remain part of the administrative-command economic system. And however much the agronomists and veterinarians were urged not to interfere in the work of a collective contract, they do not want to surrender this power' (*Arenda*, 1988, p. 97). Even if in general agreement with the need for change, people occupying positions at the top of the local hierarchy argue for gradualism and limited decentralisation, showing a distinct preference for collective and family contracts over rent. As the chairman of the state agro-industrial organization (*gosagroprom*) in Belorussia explained, 'we are in favour of large-scale production – not for a collective made up of individual renters but for collective rent as an indivisible economic organism' (*Arenda*, 1988, p.13). In Belorussia small parcels of land, he continued, could be rented to individuals but from his point of view this was 'under protest' (*Arenda*, 1988, p. 13). Caution in making changes attracts little sympathy in the media, however. As one state-farm director, accused of obstructing the formation of rented farms, put it, 'I am not an opponent of rent. I signed a contract – only I did not want to go too fast. I considered that careful preparation was necessary. Generally, the introduction of rented farms has become a campaign – reminiscent of collectivization' (*Arenda*, 1988, p. 103).

Their command over the disposition of land and other agricultural resources means that collective and state-farm managements and local bureaucrats are in a strong position to frustrate attempts on the part of farm workers to set up individual farms. It soon became clear, however, that they had allies in the peasants themselves who were reluctant to come forward to negotiate contracts. There are many reasons for this, including the expectation of stone-walling on the part of farm managements, opposition from fellow workers, a lack of confidence in the force of rental agreements and shortages of capital and appropriate technologies. As one official in Pskov oblast (in the northwest region of the European RSFSR) observed, independent farming does not immediately deliver results: 'For many farmers life is far from sweet. They don't live in well-appointed cottages, but have to put up with draughty huts. Some cowsheds have only one type of equipment – the hands of the cowherd. Anyone who hoped to make an easy rouble leaves. Those who remain need real help. Of course, some farmers are able to make repairs and to rebuild using their own savings. But the majority have only miserly funds at their disposal, while others are in debt' (*Arenda*, 1988, p. 57). These practical difficulties are especially great in the more remote farming areas of the Soviet Union, such as in the north or on loss-making collective or state farms that can do little to help aspiring independent farmers. Yet it is precisely in these more marginal situations that, according to the official view, independent farming is supposed to be concentrated (*Summary of World Broadcasts*, 26 January 1990).[1]

The practical difficulties for individual or small groups of farm workers wanting to rent their own farm means it is only the most committed who can be expected to follow this path. But according to observers such people are few and far between in the Soviet countryside. In statements that often seem to be informed by as strong a prejudice against rural people as in earlier times, the slow pace of change in the countryside has been explained by the need to combat 'the dense economic illiteracy not only of the ordinary workers, but of the specialists as well', or by 'psychological factors': rural people are said, for example, to have 'the psychology of people, who have become used over the years to working in a slapdash way, while receiving decent salaries, wages and bonuses' (*Arenda*, 1988, p.5). Apparently, rural people have forgotten the principle of 'tenacious and honest labour' and prefer to forgo the possibility of trebling their income rather than do a proper day's work (*Arenda*, 1988, p. 50). Even academician Tikhonov, otherwise a sympathetic commentator on rural life, describes the peasants' psychological state as that of the 'servile hired worker' (Tikhonov, 1989, p. 40). Unwilling to put effort into individual farming themselves, the mass of rural people, apparently, try to obstruct those who do. According to one collective-farm chairman in Orel oblast, 'Kolkhozniki intrigue against their neighbours who have negotiated rental contracts: they spoil their machinery, their crops and pour chemicals and kerosene in their milk' (*Arenda*, 1988, p.50). Similarly, Tikhonov reported that, 'members of cooperatives formed both within collective farms and outside them . . . are the butt not only of attacks from various officials but also they become outcasts in their villages'(Tikhonov, 1989, p. 40).

In the absence of appropriate social surveys, it is premature to comment on the motives said to be driving rural people in their response to the Communist Party's reform initiatives. There is little doubt, however, that the land reform has been relatively unsuccessful to date. Within the Communist Party the response of the 'right' to this has been to argue the need to refocus agricultural policy on to strictly production issues and to increase investment in intensification.[2] But for the progressive Communists the lesson of the 1980s is that the reform must be taken further and that, in particular, measures need to be introduced to lessen the ability of state- and collective-farm bosses to influence the spread of independent farms. In a decree published in June 1990 a move was made in this latter direction. Under it farm workers were given the right to leave their parent farm taking a share of its land with them, to dispose of the product of this land through whichever channels they chose and to pass their farm on to their heirs. The final arbitrators in any disputes arising over such withdrawals were to be the local Soviets, the democratically elected organs of local government, which were also vested with the power to allocate land from local reserves to other Soviet citizens who wanted to take up independent farming. By October these measures had been taken further in the Shatalin Plan, the programme for economic reform jointly initiated by Gorbachev and Yeltsin (*Perekhod k rynku*, 1990). Clearly, the accelerating political events of late summer 1990, which threatened to undermine the authority of the centre in the Soviet Union, were pushing Gorbachev further down the road to radical reform. These same political pressures, and the prospect of a winter of food rationing, led to an announcement by Gorbachev that he was prepared to have a national referendum on land privatization.[3]

The proposals contained in the Shatalin Plan reflected the political and economic preoccupations of the Soviet people in late 1990: namely, the distribution of power between the centre and the republics, and the speed of the transition to a market economy. For example, the programme emphasized that the character and pace of land reform is essentially a matter to be decided by republican governments and their local Soviets. To this end, the programme affirmed the right of local Soviets to allocate land to potential users and provided for the setting up of Land Committees in each republic responsible for re-registering land and for arbitrating land disputes. The aim of the land reallocation and registration process outlined in the programme was to transfer all agricultural land into 'the hereditary and inalienable' ownership of Soviet citizens. The implications of this measure were different for the various types of existing land holders. Collective and state farms, for example, would have their land and productive assets privatized through the allocation of shares to their members, 'private plots' would become truly private, their ownership transferred for a small sum to their current users, and individual private farms could be formed either by peasants seceding with their share of collective farm land from their parent collective or state farms or by other citizens renting land from the local Soviet's reserve. One implication of the plan was that loss-making collective and state farms would be declared bankrupt if they were unable to cover their costs. Land released by such bankruptcies was to be allocated by local authorities to the former farm members in individual rented lots. According to the programme, some 25 per cent of all the Soviet Union's collective and state farms, covering 30 per cent of its agricultural land, could expect such a fate.

Like many other radical reform proposals, the land reform contained in Shatalin's programme was soon to run into difficulties with conservatives in the Communist Party and the Soviet of People's Deputies. Gorbachev's own drift to the right late in 1990 ensured that, at least in the short term, the programme was left in a limbo; some of its measures have apparently been acted upon (for example, the task of establishing republican land committees, drawing up republican land codes and re-registering land ownership began in October 1990) but others have been shelved or watered down (such as the provision for private ownership which, if it does go ahead, will fall short of establishing a market in land). How acceptable the Shatalin proposals would have been to the diverse republican interests that by the late-1980s were coming to dominate much economic debate in the Soviet Union must remain in doubt, however. It is arguable that by this time no Moscow-inspired initiative, even one allowing for flexibility in implementation, would be acceptable to republics already set on a collision course with the centre. The Russian republic, under the presidency of Yeltsin, is a case in point. In November of 1990 the Supreme Soviet of the RSFSR introduced its own land reforms. These were clearly influenced by the Shatalin programme but also contained some novelties. For example, the right to own land in the Russian republic was given only to citizens of the republic and land sales were legalized after a period of five years (*Commersant'* 1991, p.2). The new laws reflect the current nostalgia for peasant farming among the Great Russians – the title of one of the laws is 'On peasant farms in the USSR'. The expectation is that the newly formed private farms will be approximately 50 hectares, and formed by 'peasants' seceding from their collective and state farms. Clearly, the right to secede, which under the new laws could be exercised without the consent of the parent farm, potentially puts the socialized sector in a vulnerable position. In December 1990 Yeltsin confirmed this vulnerability when he called publicly for the immediate dissolution of all collective and state farms in the Russian republic.

In the fluid political situation of 1991 it is difficult to predict the outcome of the various land reform measures proposed in the last few years; in particular, republican initiatives will depend upon political developments at the national level. It would be wrong, however, to assume that even if laws like those passed by the Russian republic were to be approved by Moscow, their implementation would be straightforward. Some of the problems that have dogged the attempts to decentralize responsibility within the collective and state farm system are likely to continue to frustrate any attempts at reform. For example, collective-farm and state-farm managers are unlikely to look favourably on measures which, taken to their conclusion, will result in the dissolution of their farms. Similarly, given current political uncertainties, no law passed at the republican level can offer a firm guarantee against reversals in the future. Farm workers are, therefore, unlikely to be much more enthusiastic about full privatization than about the previous reform initiatives. Apart from these 'human obstacles' the perennial shortage of small machinery, spare parts and credit will continue to act as a constraint on the development of peasant farming.

Notwithstanding the various problems associated with reforming the present farm system in the Soviet Union, change is inevitable if there is to be an up-turn in agricultural production. This will probably involve some diminution of the authority of collective- and state-farm managements. The pace and nature of change will vary in different regions of the Soviet Union. Furthest strides towards independent farming have been made already in the peripheral republics such as the Baltics and the Transcaucasus; in the former there is a strong tradition of independent farming and in the latter there is a strong cultural disposition towards the competitive market. In the agricultural heartland of the Soviet Union – the Ukraine, the Kuban and North Caucasus – the dismantling of collective and state farms is likely to continue to encounter strong resistance from farm managements and workers alike. However, reports from the Ukraine indicate that subdivision of state and collective farms, which had been forcibly amalgamated during Khrushchev's mergers in the late 1950s, is taking place. Elsewhere in the Russian Republic and in Belorussia, north of the good cereal lands, there is potential for rapid change: it is here that the largest number of loss-making collective and state farms is to be found. These are prime candidates to be sold off to individual citizens. However, a constraint in this region is that few people might be prepared to take up the challenge of private farming under environmentally poor conditions and inadequate infrastructure and rural services. In an attempt to revive their agricultural fortunes, collective farms in the non-black earth regions are trying to lure townspeople back into the countryside to take up individual farming (Pallot, 1990, pp. 664–7). The incentives offered include monetary grants and loans, help with equipping and stocking farms and priority access to rural services. To date such schemes have met with only marginal success. The least certain prospects for farm privatization must be in the regions of specialized cash crop agriculture and high population densities, such as in Central Asia. Developments in such regions will be determined by the extent to which current agricultural specialization continues. Improvements in efficiency on the existing large-scale cotton collectives in Uzbekistan, for example, will result in high levels of rural unemployment, but a reversion to smaller, diversified units would deprive the region of its principal source of export earnings.

Towards an agricultural market

While the establishment of a system of independent farms is proceeding only slowly in the Soviet Union, changes have been rapid in other spheres of the agricultural economy. The watershed year was 1987, when Gorbachev ordered the agricultural sector to adopt full-cost accounting, founded an agro-industrial bank, called for price reform and commissioned a new collective farm charter. The most significant change to have taken place to date is the freeing of wholesale prices from 1 January 1991. This is designed as a first step towards introducing a free market in agriculture, the sector having been identified as the one 'best prepared for the introduction of market relations' (*Perekhod k rynku,* 1990, p.166). Gorbachev has been reluctant to sanction full marketization, however; state procurements from collective and state farms in the form of the *goszakaz* remain for certain products and prices in the retail trade will continue to be subsidized. Notwithstanding these limitations, by 1990 approximately one third of consumer demand for food stuffs in the Soviet Union was being met through the free market via direct sale of produce from 'private plots', collective farm markets and consumer cooperatives.

The establishment of the agro-industrial bank also represents an important step towards the market. Its purpose is to provide credit to support the purchase by independent producers both in and outside the state system of improved machinery and farm equipment. Continuing shortages of appropriate scale technologies and spare parts may mean that the full advantages of the bank are not realized in the short term, however. Partly in response to this problem, a scheme was launched in 1989–90 whereby above-quota surpluses could be sold to the state for hard currency: the idea was that this money would be used by farms to purchase foreign farm machinery. The scheme got off to an inauspicious start, however; there were few takers for it[4] – not least because of bureaucratic delays and obstacles, and because the purchase of foreign cars proved to be more attractive to the Soviet Union's new brand of peasant farmer than buying a tractor (*Izvestiya,* 1990, p.2). Neither the organizational nor economic changes introduced by the end of 1990 had, thus, been able to make much inroad into the crisis of production and supply in Soviet agriculture. This has left the future of agriculture a live and potentially divisive issue for Gorbachev.

Social measures in the countryside

The improvement of the living and working conditions of rural people has become an integral part of 'agrarian' policy under Gorbachev. Although the annual publication of statistics in *Narodnoye Khozyaystvo* since the war has indicated an upward trend in the provision of rural services, there is no question that in much of the rural Soviet Union people are relatively deprived compared with their urban counterparts (see Chapter Two). This is particularly the case for people living in the non-black earth belt, Siberia and other marginal regions, and for specific categories of rural people, such as the elderly and the physically and mentally handicapped. One consequence of the low standard of living in rural areas is that rural-urban migration has become a flood.[5] Out-migration has left some places badly depopulated. Table 5.3 shows that the central regions of the European Soviet Union, where out-migration has been most intense, have been experiencing negative growth rates for more than a decade. The first regions to register, in the 1960s, such negative growth rates in rural districts were Novgorod, Pskov, Smolensk,

Table 5.3 *Rural population change in the major economic regions of the USSR, 1951–87*

Republic	Economic region	Per cent change
RSFSR		–32.6
	Northern	–28.0
	North West	–42.0
	Central	–54.0
	Volga-Vyatka	–51.7
	C. Chernosem	–50.2
	Volga	–35.7
	N. Caucasus	+13.2
	Urals	–28.0
	W. Siberia	–28.3
	E. Siberia	–10.0
	Far East	+27.0
UKRAINE		–28.3
	Donets-Dnieper	–39.9
	South-West	–26.8
	South	–2.6
BELORUSSIA		–40.4
LITHUANIA		–32.0
LATVIA		–25.7
ESTONIA		–21.0
MOLDAVIA		+13.3
GEORGIA		+5.5
AZERBAYDZHAN		+94.2
ARMENIA		+43.0
UZBEKISTAN		+147.8
KIRGIZIYA		+98.0
TADZHIKISTAN		+183.5
TURKMENISTAN		+132.8
KAZAKHSTAN		+64.6

Source: Naseleniye SSSR (1987) Finansy i Statistika, Moscow.

Ivanovo, Kostroma, Ryazan, Tula and Yaroslavl. By the late 1970s the number of such regions had risen to 15 and today there is negative growth in 30 of the 38 regions in the European part of the Russian Republic, 15 of the 25 regions in the Ukraine and 4 of the 6 in Belorussia (See Map 5.1). Among these regions some stand out for having experienced particularly sharp declines: for example, in the centre of European Russia the rural population in Kalinin region fell 63 per cent between 1951 and 1987, in Ryazan by 60.1 per cent and in Bryansk by 55.4 per cent. In total, the rural population of the central regions of the European Soviet Union has declined by nearly 10 million in the past 25 years (Rybakovskii and Tarasov, 1989, p. 73). Such levels of population decline have had an impact upon the viability of Russia's villages. As the results of the 1989 census show, there is a process of progressive settlement concentration taking place in the Soviet Union which has reduced the total number of villages by 51,000 in the last 10 years (Table 5.4). It is the smallest and more remote settlements with ageing populations that are most vulnerable to desertion and which will disappear from the map of the rural Soviet Union if measures are not taken to stabilize their populations. A.M. Yemel'yanov voiced a growing concern when he said: 'It is essential that we do everything possible to save the village. And it is necessary to start with people – to create normal

Table 5.4 *Settlement size distribution, 1979 and 1989*

Settlements by population size (thousands)	Number of settlements		% distributions of rural population by settlement size	
	1979	1989	1979	1989
< 10	105	99	0.4	0.4
11–50	86	71	2.3	1.9
51–100	42	31	3.1	2.3
101–200	42	32	6.2	4.7
201–500	53	45	17.3	14.9
501–1000	30	28	21.5	20.4
> 1000	25	26	49.2	55.4
Total	383	332	100	100

Source: 1989 census.

conditions for life in the village, closer to conditions in the towns. And we are referring here not only to raising wages. Today the question is much broader – it is about guaranteeing social equality between rural and urban dwellers. I am referring here to housing conditions, retail trade, cultural facilities and medical care, road and transport provision, in other words everything a person needs' (*Moskovskii Komsomolets,* 1989, p. 2).

Rural out-migration is not confined to the European part of the Soviet Union. The population growth shown in Table 5.3 for the Far East and Siberia in fact disguises high rates of population turnover which causes labour problems for agriculture. In Central Asian the problem is different again. High rural birth rates in the four republics have led to overpopulation and poverty. In this respect they resemble typical third world countries.

Gorbachev's interest in raising rural living standards is long overdue. On 5 April 1989, less than a month after the Communist Party plenum at which the go-ahead was given for independent farming, the Communist Party unveiled a new programme for reviving rural society (*Pravda,* 5 April 1989). In it there is a pledge that by the end of the thirteenth five-year-plan (1995) all rural districts will be provided with a basic service infrastructure that will include running water, electricity, a gas system and road network, and that schools, kindergartens, clubs, houses of culture and retail outlets will be built on all collective and state farms. By 1998 every rural inhabitant is supposed to have twenty-one square metres of living space. Considering the existing low level of service provision in rural regions of the Soviet Union, the estimated cost is inevitably high – some 121 billion roubles, which, according to the new program, should be raised from a variety of sources including central and republican funds, banks and private individuals. The attempt to shift much of the burden of rural service provision from central to local authority and private capital follows the established practice of the perestroyka era in the Soviet Union, but success is far from guaranteed. Rural servicing will have to compete with other calls on local and private capital but since its returns will mainly be felt in the long term it is unlikely to receive a high priority. Central support will therefore be crucial to the success of the social program. The decree does indicate that some additional central funding will be available to support local building programs but it skates over what proportion of the total financing is projected to come from public, private and local authority sources. In the current state of economic crisis it is difficult to see how an ambitious programme of rural building and service provision can get off the ground in other than the most prosperous regions of the country.

Privatizing the rural housing stock

One sphere of construction in which big strides forward can be expected in rural areas is house building. In line with the move towards private enterprise in production, private house building is being actively encouraged. This reverses the trend in the post-war years when the proportion of new housing stock put up in rural areas by collective and state farms or local authorities rose from under 50 per cent in the 1960s to 60 per cent–70 per cent at the end of the 1970s and beginning of the 1980s (*Naseleniye SSSR*, 1987, pp. 8–95). Much of this housing was in the form of blocks of flats, which were preferred for cost reasons and because they helped to create the impression of 'urbs in rure', but of notoriously poor quality. For example, of the new rural dwellings built in the Russian republic in 1985 only 30 per cent were fully provisioned with running water, sewage and central heating systems, while 42 per cent had none of these. In the Ukraine and Central Asia the record was even worse, as Table 5.5 shows.

It has been calculated that in order to expand and up-grade the rural housing stock the current annual expenditure on house building will have to be doubled to twenty milliard roubles (*Eknomicheskaya gazeta*, 1989, p. 7). Individuals will be encouraged to take out loans to build and there are especially favourable terms for return migrants and people negotiating rental contracts with farms. The hope clearly is that the ability to own one's home will be a big incentive for people not to leave the countryside and to repopulate outlying settlements. *Sel'skaya Zhizn'*, the Soviet Union's rural daily newspaper, now carries many articles making this point. For example, one correspondent reporting from a village of newly built private houses in the central part of European Russia assured readers that, 'people do not want to leave these villages . . . there is no shortage of labour. There are lots of young people here . . . it's true what they say: a private home is your own – you don't get up and leave your own home' (*Sel'skaya zhizn'*, 1988, p. 1). But there are hints that the magic does not work with all generations; it has been found, for example, that the young are less interested in building their own home than using their savings to buy a car. This tendency has been blamed on the loss by young people of their 'rural roots', an over-reliance on the willingness of the state to provide and a general softening of character (*Sel'skaya zhizn'*, 1988, p. 2). Another problem that might stand in the way of private house-building is the shortage of appropriate building materials. The emphasis in previous years on high-rise, prefabricated housing resulted in the closure of the small rural brick factories and

Table 5.5 *The degree of amenity provided to new rural housing in 1985 (by republic)*

	Percentage of new houses with:		
	sewage system, running water and central heating	partial provision	no provision
RSFSR	30	28	42
UKRAINE	23	34	43
BELORUSSIA	42	23	35
LITHUANIA	86	13	1
LATVIA	79	16	5
ESTONIA	92	8	–
MOLDAVIA	62	26	12
GEORGIA	100	–	–
AZERBAYDZHAN	–	40	60
ARMENIA	69	18	13
UZBEKISTAN	23	57	20
KIRGIZIYA	24	30	46
TADZHIKISTAN	4	24	72
TURKMENISTAN	28	19	53
KAZAKHSTAN	14	23	63

Source: Vestnik statistiki (1987) 5, p. 68.

quarries needed to supply the local building industries. Building cooperatives and private enterprise are expected to fill this gap.

Conclusions

Rural policy in the Soviet Union is almost synonymous with agricultural policy; ultimately, reforms in the countryside, whether to do with the extension of private property, the improvement of rural services or with other issues, such as the reorganization of local planning mechanisms, are designed to improve the Soviet Union's agricultural economy. This emphasis in Soviet policy is understandable since, as pointed out at the beginning of this chapter, the success of perestroyka depends in large measure on improvements taking place in agriculture. Failure to make improvements could have serious consequences for the economy, already in crisis, and for political stability. Food shortages touch the daily life of every Soviet citizen and are a potential source of social unrest. The food question also links in with the, as yet, unresolved problem of the relationship between the Soviet Union's republics and regions. In conditions of general food shortage, traditional surplus producers may be more reluctant to export than in the past. The Baltic States have already put bans on the export of certain food items to other republics in the Soviet Union, and have even sought to export food to the West for hard currency.

At present much hope is being placed on the introduction of private farms and the market to solve the agricultural problem. For an increasing number of people in the Soviet Union, a precondition for such changes is the dissolution of the existing collective- and state-farm system along with the associated planning bureaucracy. The vision of the future rural landscape in the Soviet Union is very different from what exists today: small hamlets will exist side-by-side with large villages, both equally well-serviced, and diversified peasant farms will stand side-by-side with medium to large intensive farms run by individuals or small groups of cooperators. Presumably an efficient system of roads, grain elevators and other storage facilities and of market towns will help convey the product of this new, efficient farm landscape to the large conurbations. The vision is an attractive one, especially for a people who have suffered intermittent food shortages for half a century, but the problems in attaining it are extremely large. Whether the present or some future Soviet leadership will be able to develop a program that will enable the transition from the old to the new to be made without, in the process, incurring high social and economic cost must remain in doubt. One certainty is that an immediate solution to the Soviet Union's agricultural problem is not in sight.

Notes

1. That loss-making collective farms are the prime candidates for dissolution has become an accepted element of current agrarian reform. This was spelled out by Stroyev, central committee secretary in charge of agriculture, in an interview with *Pravda* on 27 December 1989. However, it can be questioned whether weak *kolkhoxy* are, in fact, the best soil on which to experiment in independent farming; see, for example, Macey, 1990, p.12.
2. Ligachev, former Politburo conservative and opponent of Gorbachev, consistently argued for more investment in agriculture and spearheaded criticisms of the land reform.
3. The promise of a national referendum on private ownership of land has been made by Gorbachev on several occasions. As of February 1991 it had not been carried out.
4. In 1989 only 1 per cent of all agricultural marketings were paid for in hard currency and farms made only 19 million hard currency roubles from the scheme.
5. There is some discussion among Soviet geographers about the relationship betwen rural depopulation, the level of rural services and urbanization. The 'majority' view is that the low standard of living in rural areas has acted as a push factor to migrants, but Grigorii Ioffe (1986) maintains that its influence has been exaggerated and much rural-urban migration in the Soviet Union can be explained by the 'normal' redistribution of population associated with urbanization.

Suggested reading

Lewin, M. (1968), *Russian Peasants and Soviet Power: A Study of Collectivization,* Allen and Unwin, London.

Medvedev Z. (1987), *Soviet Agriculture,* W.W. Norton and Company, New York and London.
Pallot J. (1988) 'The USSR' in Cloke P. (ed.) *Policies and Plans for Rural People: an International Perspective,* Allen & Unwin, London.
Pallot J. (1990) 'Rural Depopulation and the restoration of the Russian village under Gorbachev', *Soviet Studies,* 42, (4), pp. 655–74.
Stuart R.C. (1972), *The Collective Farm in Soviet Agriculture,* Lexington Books, Massachusets.
Wadekin, K.-E. (ed.) (1989), *Communist Agriculture: Farming in the Soviet Union and Eastern Europe,* Routledge, London.

References

Arenda – arkhimedov rychag (1988), Pravda, Moscow.
Commersant (Weekly Business Paper, in English and Russian).
Ekonomicheskaya gazeta (Weekly Economics Paper in Russian).
Ioffe, G. (1986), *Nechernozem'e: sotsial'naya geografia i khozyaistvo,* Nayka, Moscow.
Izvestiya (1990), 'krest'yane vykhodyat na volyntnyi rynok', 27 October, p.2.
Kingston-Mann, E. (1985), *Lenin and the Problem of Marxist Peasant Revolution,* Oxford University Press, New York.
Macey, D.A.J. (1990), 'Gorbachev and Stolypin: Soviet agrarian reform in historical perspective', *Comparative Economic Studies,* 32, (2), pp. 7–28.
Medvedev Z. (1987), *Soviet Agriculture,* W.W. Norton & Co., New York and London.
Moskovskii komsomolets, (Moscow Daily Newspaper in Russian).
Naseleniye SSSR, 1987, Goskomstat, Moscow.
Pallot, J. (1979), 'Rural settlement planning in the Soviet Union', *Soviet Studies,* 31, (2), pp. 214–30.
Pallot, J. (1987), 'Continuity and change in village planning from the 18th century', in Holzner, L. and Knapp, J.M. (eds) *Soviet Geography in our Time: a festschrift for Paul E. Lydolph,* Wisconsin, pp. 319–50.
Pallot, J. (1988), 'The USSR', in Cloke, P. (ed.) *Policies and plans for rural people: an international perspective,* Allen & Unwin, London, pp. 120–41.
Pallot, J. (1990), 'Rural depopulation and the restoration of the Russian village under Gorbachev', *Soviet Studies,* 42, (4), pp. 655–74.
Perekhod k rynku. Kontseptsiya i programma (1990), Arkhangel'skoye, Moscow.
Pravda (1989), 'O programme sotsial'nogo razvitiya sela', 5 April, pp. 1–2.
Rybakovskii, L.L. and Tarasov, N.V. (1989), 'Sovremennyye problemy migratsii naseleniya' *Istoriya SSSR* (2).
Sel'skaya zhizn' (Daily Agricultural Newspaper in Russian).
Summary of World Broadcasts (Daily summary of Russian Language Broadcasts).
Tikhonov, V.A. (1989), 'Uroki proshlogo i nastoyashchego', in *Glasnost'*. Nasushchnyye voprosy i neobkhodimyye otvety, politizdat, Moscow, pp. 31–42.
Wadekin, K.E. (1973), *The Private Sector in Soviet Agriculture,* University of California Press, Berkeley.

6

Perestroyka and Siberia: frontier resource development

Gary Hausladen[1]

Introduction

For over four centuries, Siberia has served as a vast resource frontier for the Russian heartland. Since the Cossacks first extended Tsarist control into 'the land beyond the Urals', Siberia has provided a bounty of resources to fill the coffers and stimulate the growth of the national economy. Siberia has also served as a region of direct confrontation with and access to the Orient, as a bridge to East Asia and the Pacific Rim. It is a land whose potential riches and strategic location have whetted the appetites of Russian and Soviet leaders for centuries, as they have attempted to overcome the seemingly insurmountable obstacles to its development and integration.[2]

This vast territory called Siberia, which refers to all three of the eastern economic regions of the RSFSR (Map 6.1), is more than merely a land of mystery and allure. It is also one of the keys to understanding the growth potential of the Soviet economy, especially within the framework of Gorbachev's reform programs. There is an explicitly geographical dimension to these reforms and Siberia figures prominently in that dimension. In fact, some have suggested that perestroyka will produce a new phase in the relationship between the Siberian periphery and the Russian heartland as the Soviet national economy undergoes basic structural changes. Given recent political challenges to Soviet hegemony over the RSFSR, one must further question whether the future role of Siberia will evolve as part of an All-Union or a Russian economy.

Whether Siberia's role within the Soviet and/or Russian economy is about to take on new dimensions, and how these new dimensions will manifest themselves, may be questioned. What is clear, however, is that there has been both continuity and change in the basic relationship between the European core and the Siberian resource periphery. Siberia has played an important role in the development of the Russian and Soviet political economies and it will continue to do so in the future. Siberia, and its relationship to the Russian heartland, may well serve as a barometer of the degree to which economic reforms of any kind succeed.

Today, the basic characteristics of Siberia's role within the Soviet political economy remain its role as a resource periphery and as a gateway to the Pacific Asian region. In this regard, it is necessary to balance local with national and global factors to appreciate the causes for Siberian development. To the present, the 'Siberian experience' has been treated as a singular phenomenon, understandable only within its Russian and Soviet context. There are characteristics of the Siberian experience that are singularly Siberian. At the same time, these kinds of analyses have been blind to those aspects of Siberian development that have been in response to global processes and the changing role of Russian/Soviet integration into the world economy. Both perspectives must be balanced to understand fully Siberia's changing role as a resource frontier to the Russian Empire and the Soviet state. Only when we put Russian industrialization during the Stalin and post-Stalin years in the context of inter-state competition, exacerbated by ideological considerations, can we understand continued development of resources at all costs in regions of low-investment productivity. Especially in the context of perestroyka, both local and global processes will influence Siberia's changing role in the Soviet economy.

The legacies of tsars and commissars

In Siberia, as elsewhere, Soviet political and economic geographies have been built upon tsarist foundations. It is only when we understand these past geographies that we can better appreciate contemporary patterns of location and interaction. And it is only when we understand contemporary patterns that we can better assess potential ramifications of reform policies. Quite clearly, this evolution has many components, some of which stress continuity, some of which stress change. From the very beginning, the exploitation and development of Siberia was partially in response to global processes in the evolution of an increasingly integrated world economy. This also

[1]The work for this chapter was supported by funding provided by the National Council for Soviet and East European Research to which the author is most grateful.

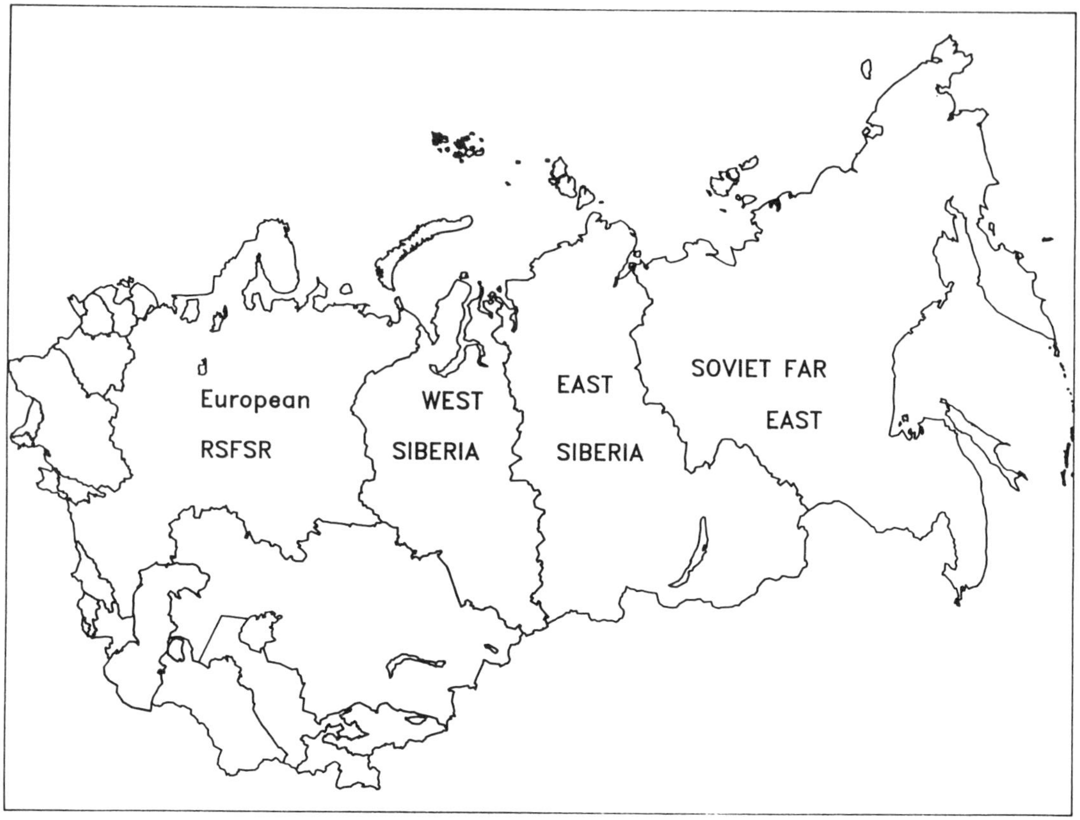

Map 6.1 *Soviet Siberia*

helps to explain the historical balance between continuity and change in the basic nature of the relationship between Siberia and its various subregions, and the Russian core.

The tsarist legacy: foundations for Soviet colonialism

By the end of the tsarist period, it is possible to characterize the relationship between the Russian heartland and Siberia as a classic core-periphery relationship. Although the inclusion of Siberia into the Russian state was facilitated by the fact that it was contiguous, the basic nature of the relationship remained colonial. Many of the characteristics of this relationship changed over three centuries, but the basic dependency between core and periphery remained firmly entrenched. Primary goods, most importantly agricultural products, furs, fish, and, to a lesser extent, minerals, provided the economic rationale for Russian interest in the region, as did access to the Pacific eventually afforded by the Trans-Siberian Railroad. At the same time, primitive methods of resource extraction and industrialization kept the eastern regions heavily dependent upon European Russia and foreign countries for manufactured goods (Lyashchenko, 1949, pp. 584–604).

For the most part, early twentieth century Siberia, specifically West Siberia, served as an agricultural colony, which supplied grain and dairy products to Europe and East Siberia. Furs for European Russia and Europe still retained importance as a source of national wealth, as did certain metals, especially gold, silver, copper and coal. Yet lack of technology in extraction and production limited the degree to which these mineral resources stimulated growth and added to the national wealth. Those enterprises that succeeded were generally financed with foreign capital, as in the case of gold production in regions around the Lena, Vitim, Zeya and Bureya rivers, or were under the direct control of the tsar, as in the case of the Altay and Nerchinsk gold and silver regions

(*Aziatskaya Rossiya,* 1974, Vol. 1, pp. 388–439, and Vol. 2, pp. 182–7). Thus, on the eve of revolution, the economic geography of Siberia already held the seeds of growth for economic integration into the national economy. To that time, however, the infrastructure had not been provided to take advantage of its relatively-untapped wealth.

The Revolution and subsequent establishment of the Soviet Union did little to change the basic nature of this relationship. In many respects, the ministerial model of control and decision-making within the Soviet economy paralleled the tsarist system of centralized control. When we add to the political-economic structure the adversities of civil war, collectivization and world war, it is not surprising that many of the ideologically-bound goals for diversified Siberian development did not receive serious attention during the first four decades of Soviet rule.

The Stalinist legacy: eastward and extensive growth strategies

The economic history of the Soviet Union has been defined by the decision of Stalin to pursue rapid industrialization in an attempt to modernize and 'catch the West'. Stalin made it clear that competition with the capitalist countries played no small role in that decision.

> We have lagged fifty to a hundred years behind the leading countries. We must cover this distance in ten years. Either we do, or they crush us (as quoted in Hutchings, 1982, p. 48).

This approach has led to unequal development for different sectors of the economy; heavy industry and the military have benefited at the expense of light and consumer industries. This strategy had specific implications for Siberian development. For Siberia, rapid industrialization meant that an increasing importance was tied to energy and mineral resource extraction, previously underutilized. Now the industrial resources of Siberia were used to stimulate industrial development in the European regions of the country. First they supplemented European supplies; later they replaced them. The economic value of West Siberia was noted by Lenin immediately after the Revolution and incorporated into the first five-year plan (FYP) in 1928–9, wherein the decision was made to develop metallurgy based on the supply of iron ore from the Urals and coal from the Kuznetsk basin (Kuzbas).

Greater emphasis on eastern development was well underway by the Second World War, as witnessed by increased industrial production, the establishment of 33 new cities (between 1928 and 1941) and increased population. From 1938–40 alone, nearly 600,000 people migrated to Siberia and the Far East (Alekseyev and Isupov, 1986, p. 50). The Nazi invasion of the Soviet Union in 1941, however, greatly increased the pace of Stalin's shifting of industrial production capacity eastward. The Nazi occupation of the western regions of the Soviet Union forced a massive migration of factories and labour. From 1941 to 1942, nearly 1,600 large-scale factories were completely evacuated from the west, as were millions of workers. Of these 244 were relocated in West Siberia and 87 were relocated in East Siberia (Dokuchayev and Kozybayev, 1987, p. 82). Additionally, over 2,000 new factories were constructed in Siberia and Kazakhstan during the war (de Souza, 1989, p. 71). As a result, from 1940 to 1945, production of heavy industry almost tripled in West Siberia and increased by nearly 40 per cent in East Siberia (Alekseyev and Isupov, 1986, p. 64).

An ancillary result of these policies was the stimulus they gave to urbanization in the region. The process of Siberian migration during the war had several components. First, millions of Siberians left Siberia to fight in the war. Most were men from the countryside. To provide the necessary labour for the increased level of industrialization, large numbers of rural Siberians as well as European urban dwellers migrated to Siberian cities. Although the total population of Siberia decreased by 1.5 million during the war, there was actually an increase in urban dwellers of nearly 700 thousand (Alekseyev and Isupov, 1986, p. 194). In fact, the migration of labour and industry during the Second World War played no small role in the urbanization of Siberia. This process accelerated trends already underway. In 1929, on the eve of rapid industrialization, only 14 per cent of the population of Siberia lived in cities; this proportion had increased to 31 per cent in 1939 just prior to the war. By 1945, at the end of the war, 43 per cent of the population lived in cities (Isupov, 1987, pp. 35 and 39).

On the one hand, shifting industrial capacity eastward resulted in increased investment in Siberian development. Yet the nature of the development reinforced Siberia's dependency on European Russia and hampered its ability to produce a multi-dimensional economic geography. The overall goal of these policies was not to benefit Siberia but to replace and augment the nation's productive strength. Although Stalin sought to create a self-sufficient Soviet economy, isolated from the capitalist world economy, it was, in fact, processes in the world economy that greatly influenced Soviet policies toward Siberia – competition with the West, which stimulated 'superindustrialization', and the Second World War, which accelerated the shift in industrial capacity eastward. By the end of the Second World War, Siberia's value economically and strategically had been clearly demonstrated and its integration into the national economy had become a priority.

The post-war period saw a dramatic but short-lived shift in priorities as the Soviet Union sought to reconstruct and recover from the devastation. This meant major investment in the western regions of the country, which had suffered incredible losses during the war. For the interim, Siberian development was put on hold. Yet, with few exceptions, the post-Stalin era proved to be no different. Although plans suggested regional equity and improvements in the levels of development in outlying regions, the actual character of investment into the region remained highly concentrated and localized, still emphasizing the importance of resource extraction, transportation links with the Pacific and military security.

The Khrushchev and Brezhnev legacies: Siberian mega-projects

One aspect of Siberian development that evolved under Khrushchev and Brezhnev was the approval of mega-projects – gigantic economic ventures that required large allocations of capital resources, with questionable levels of return to the national economy. The first of these were the huge hydroelectric power projects begun the 1950s. Constructions of these projects lasted thirty years into the 1980s. Major hydro power stations have been constructed, or are under construction, along the Angara River, at Bratsk, Ust'-Ilimsk and Boguchany; along the Yenisey, at Krasnoyarsk and Sayan-Shushenskoye; and along the Zeya, Bureya, Vilyuy and Kolyma rivers in the Far East (CIA, 1985, pp. 50–1). Each has been responsible for the establishment of high-energy consumption production, particularly aluminum. Yet the overall economic rationale may be questioned.

> It may well be that the whole policy of construction of gigantic Siberian hydroelectric power stations was a fantastically costly mistake, undertaken without adequate analysis and in ignorance of the energy crisis which the European USSR began to face from the mid-1950s onwards (Schiffer, 1989, p. 24).

Despite controversy within the Soviet Union over the advisability of hydroelectric projects in Siberia and the Far East, the 1970s saw further implementation of mega-projects, particularly the Baykal-Amur Mainline (BAM) and Territorial-Production Complexes (TPCs). Although they marked a change in Soviet strategies, they were reminiscent of similar tsarist mega-projects. In fact, BAM is often referred to as the second Trans-Siberian. The rationale for BAM was clearly in keeping with Siberia's role in the national economy. Economically, BAM was intended to open new resource regions for exploitation, for both domestic and foreign markets, and provide better access to Pacific Asia. Strategically, BAM was intended to provide a more secure route for the movement of troops and equipment *vis-à-vis* China. Both economic and strategic considerations thus played important roles in implementing a mega-project, which, like the hydro power stations, was questionable with respect to economic efficiency.

The 1970s also saw the implementation, at least in theory, of TPCs (see de Souza, 1989). Although not as 'concrete' as hydro stations and rail lines, they represent another kind of approach to Siberian economic development, at least in commitment. It is not yet clear the extent to which they are rhetoric and/or reality. It is important to point out that 'the complexes are formed to solve a national economic problem or, ultimately, make possible a future development of the national growth potential' (de Souza, 1989, p. 92). This definition could be applied to almost any economic region, with two exceptions – they are comprehensively planned, and they are designed to fulfill national, as opposed to regional or local, needs.

For Siberia and the Far East, five TPCs have been officially identified in FYP statements: the West Siberian, Kansk-Achinsk (KATEK), Sayan-Shushenskoye, Bratsk-Ust'-Ilimsk and South Yakutian. An additional eight have been planned for development along the BAM, all of which have apparently been put on hold. With the exception of the West Siberian TPC, all these are based on the production of hydroelectric power for energy-intensive industry. In the Bratsk-Ust'-Ilimsk TPC, this power is used for the production of aluminum, timber products and iron ore; in the KATEK, for aluminum, metallurgy and machine building; in the Sayan-Shushenskoye TPC, for aluminum and transportation equipment, primarily for BAM; and in South Yakutiya, for coal, primarily for export to Japan. West Siberia, as one would expect, relies primarily on thermal power for the production of oil and gas products, for which it is the Soviet Union's leading producer.

These mega-projects notwithstanding, basic approaches toward the exploitation and development of Siberia and the Far East remain constant. Hydro stations, BAM and TPCs are designed primarily to promote the priority development of natural resources for use by the national economy. The 'push' toward Siberia under Stalin as a result of superindustrialization and the Second World War, and the implementation of mega-projects throughout the post-Stalin period, resulted from continuing competition with the West, the economic growth of various Pacific countries, coupled with depletion of resources in the western regions of the country, and the need to secure the Soviet presence in Asia during the post-war period.

It is important, however, not to confuse these mega-projects with a dramatic shift in national commitment to Siberian development. Jonathan Schiffer, in his detailed analysis of Soviet economic policy toward Siberia, warns of the pitfalls of confusing rhetoric for reality. Statements of top government officials and inclusion in plan guidelines do not ensure implementation. In fact, for the post-Stalin period, he shows that Siberia's share in state cooperative investments remained relatively stable. Its share accounted for a high of 18 per cent of total national capital investments during the war years (1941–5) and dropped to a low of 13.5 per cent in the immediate post-war FYP (1946–50). Its share rebounded to 15.6 per cent in the next FYP (1951–5) and remained at just over 16 per cent from then until the mid-1970s (Schiffer, 1989, p. 29). But it increased in the late 1970s and early 1980s because of the oil and gas industry.

Siberia as periphery: on the eve of perestroyka

By the 1980s, Siberia's allure remained primarily economic and strategic. Its economic lure focused on industrial and energy resources, fish and timber products and access to Pacific Asia. Its strategic lure also focused on access to Pacific Asia, giving the Soviet Union a presence in East Asia. Yet because of its size and relative location, these benefits were not received without a price. The disadvantages of Siberian development and integration include costs associated with control and defence and problems associated with resource, extraction and transportation to regions of use, primarily in the west and overseas.

Treasure trove of resources

In value and importance, the 'soft gold' of the Cossacks has been replaced by the 'black gold' of the commissars. Yet the treasures of Siberia are diverse and dispersed, ranging from oil and gas in northern West Siberia to timber and fish in the southern Soviet Far East. For example, from the Kuzbas in the mid-1980s, the Soviet Union received 8 per cent of its steel and 20 per cent of its coal; from West Siberia, it received 65 per cent of its oil and 57 per cent of its natural gas; from Siberia and the Far East, 37 per cent of its sawn materials; and from West Siberia, 9 per cent of its grain (Molodenkov, 1987, p. 6).

In this regard, it is useful to identify several (of numerous) Siberias and to characterize briefly their value to the national economy (Map 6.2). In West Siberia these include the Ob'-Irtysh and

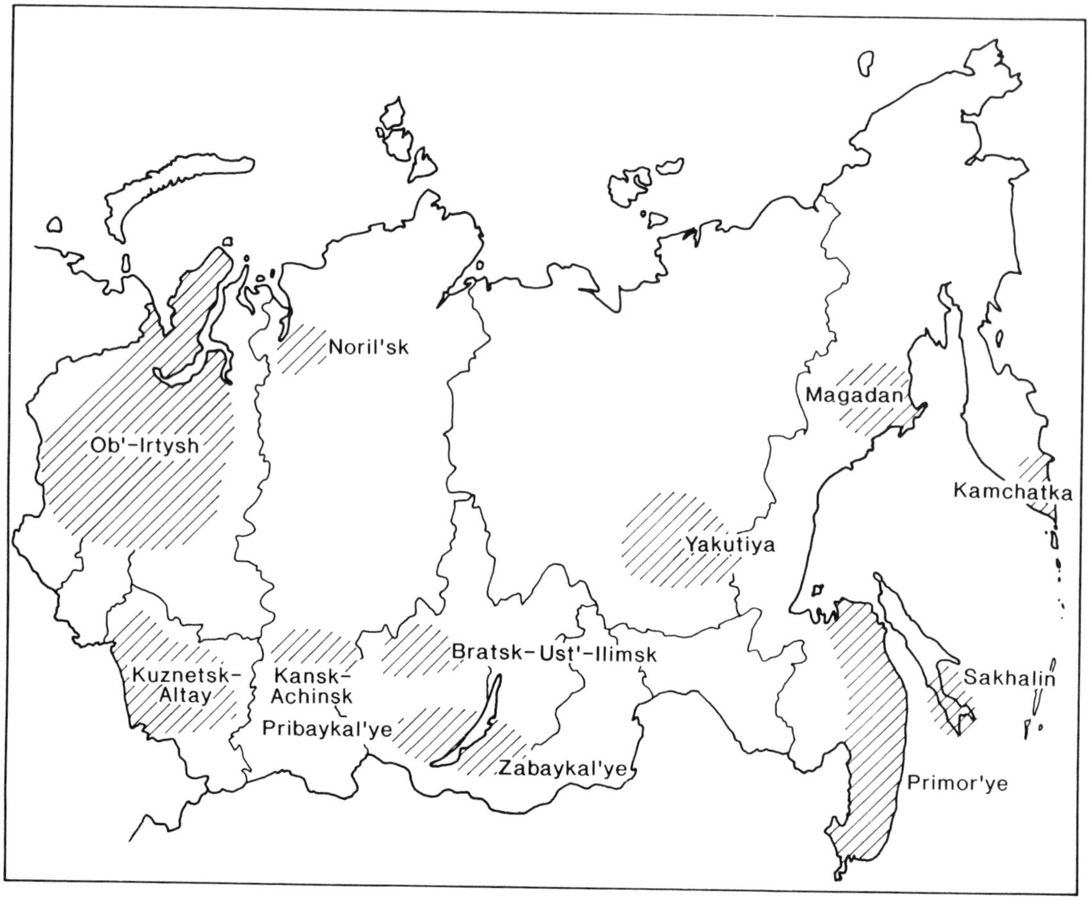

Map 6.2 *Economic regionalization of Soviet Siberia*

Kuznetsk-Altay regions; in East Siberia the Noril'sk, Bratsk-Ust'-Ilimsk, Kansk-Achinsk, Pribaykal'ye and Zabaykal'ye regions; and in the Far East the Yakutiya, Magadan, Kamchatka, Sakhalin and southern Far East regions. It is not by coincidence that several of these have already been identified as TPCs within the context of Soviet planning strategies.

In West Siberia, the Ob'-Irtysh region includes Tyumen', Omsk and Tomsk oblasts. Key centers in the region include the cities of Surgut, Nizhnevartovsk and Urengoy. For the national economy, this region is synonymous with the production of oil and gas. It also produces additional petroleum products, agricultural and fish products and has some machine building, primarily for local use. In southern West Siberia, the Kuznetsk-Altay region includes Altay kray, and Kemerovo and Novosibirsk oblasts. Major industrial cities of the region include Novosibirsk, Barnaul and Kemerovo. For the national economy, this region is a major supplier of coal. It also produces iron-ore, aluminum, machinery, foodstuffs and timber products.

In the far northern reaches of East Siberia, the region around Noril'sk provides valuable ferrous and non-ferrous metals. Of importance are nickel, cobalt, lead, zinc and copper. The Bratsk-Ust'-Ilimsk region, to the area north of Irkutsk, based on hydropower from dams at Bratsk and Ust'-Ilimsk, produces timber products, iron ore and aluminum for the national economy. The Kansk-Achinsk (KATEK) region includes southern Krasnoyarsk oblast. The key industrial city in the region is Krasnoyarsk. For the national economy, this region is a major supplier of coal, iron ore, timber products and aluminum. It also produces machinery, primarily for local use.

Pribaykal'ye is basically Irkutsk oblast. It includes hydropower from the Angaro-Usol'ye node. Irkutsk is the dominant city of the region. For the national economy, this region contains important sources of molybdenum, mica, coal and timber. It also produces machinery and aluminum. Zabaykal'ye includes Chita and Buryat ASSRs to the southeast of Lake Baykal. Key cities in the region are Chita and Ulan-Ude. For the national economy this region produces gold, timber products and non-ferrous metals.

In the Far East, Yakutiya includes the region around Yakutsk and southern Yakutiya, around Neryungri. For the national economy it supplies coal, diamonds and iron ore. It also produces agricultural products for local consumption. In the Pacific Far East there are three, relatively isolated sub-regions: Magadan, Kamchatka and Sakhalin. Each is located near an important port facility: Magadan, Petropavlovsk-Kamchatskiy and Yuzhno-Sakhalinsk. For the national economy Magadan is a major source of gold; Kamchatka supplies coal and Sakhalin is beginning to develop its deposits of oil and gas. Each also provides access to the Pacific. The final region includes most of the southern Pacific Far East, to include Khabarovsk and Maritime krays and Amur oblast. The major cities of the region are primarily ports – Sovetskaya Gavan', Khabarovsk (inland port), Nakhodka and Vladivostok as well as the region's major industrial center, Komsomol'sk-na-Amure. For the national economy, this region is a major supplier of fish, timber products, coal and non-ferrous metals. It also produces machinery, especially those kinds associated with ship building and repair. In addition to the natural resources, these ports and entrepôts provide greater access to the Pacific for both economic and military purposes.

Access to Pacific Asia: the economic dimension

For centuries, tsars and commissars have appreciated the benefits to be gained from integration into the world economy. Siberia has always played a key role in this integration. In the sixteenth century, Russia supplied furs to the capitals of Europe; later, silver and gold were important exports; these were replaced in importance in the twentieth century by energy and industrial resources for world markets.

During the 1960s and 1970s, the complementary relationship between resource-rich, technologically-poor Siberia and resource-poor, technologically-rich Japan served as an example of the potential for direct interaction between the eastern regions of the Soviet Union and the developed and developing countries of the Pacific Rim. For a variety of reasons, Soviet-Japanese relations never lived up to their potential. Within the context of new economic programs, this potential may become unlimited. To quote Gorbachev, 'Soviet-Japanese relations have a contradictory past, a very promising future, and a relatively barren present' (*CDSP*, 30 January 1991, p. 27).

Currently, the nature of limited Soviet-Japanese trade provides only glimpses of the possibilities. Soviet imports from Japan are dominated by manufactured goods and high technology, especially machinery and equipment, rolled steel and pipes. Soviet exports to Japan are dominated by natural resources, especially timber, coal, petroleum and petroleum products. Yet total Soviet-Japanese trade accounts for less than 3 per cent of the total Soviet trade turnover and only 2 per cent of the total Japanese trade turnover (*Vneshniye*, 1988, pp. 9–14; Ogawa, 1987, p. 159).

This scenario also applies to Soviet trade relations with the other capitalist countries of Pacific Asia. Total trade with Malaysia, Thailand, Singapore, Indonesia and the Philippines combined represents less than 1 per cent of total Soviet trade turnover (*Vneshniye*, 1988, pp. 9–14). This does not include trade with South Korea, which approached $US 280 million in 1988 (Lho, 1989, p. 1162) and the possibility of trade with Taiwan, with whom trade relations, according to Gorbachev's Vladivostok speech, will now be possible. When we combine Soviet trade with all capitalist countries of the Pacific Rim, to include Japan, Canada, the United States, Australia and New Zealand, the total still represents less than 5 per cent of Soviet trade turnover (*Vneshniye*, 1988, pp. 9-14). The bases for increased trade and economic relations are set. The degree to which the true potential of Pacific Rim trade will be realized in the near future, however, is still in serious doubt (see Bradshaw, 1988).

Access to Pacific Asia: the strategic dimension

The distinction between the economic and strategic dimensions of Soviet access to the Pacific is somewhat artificial. Because one of the functions of the modern nation-state is to secure and protect national economic interests throughout the world, economic and strategic factors are closely interrelated. In capitalist countries, this interrelatedness is referred to as the military-industrial complex, implying interaction between the government and private companies, most importantly multi-nationals, in domestic and international affairs. In the Soviet case, where the state controls both components of the military-industrial complex, the distinctions are even more obscure. The primary rationale for a Soviet presence in Siberia and the Far East is economic. The military, as one means of enforcing government policy, is there basically to protect its economic interests – to defend borders, to secure transport routes into and out of the region and to encourage and support friendly, or at least neutral, buffer states on its periphery.

The nature of the Soviet military presence in Siberia and the Far East changed dramatically during the Brezhnev years. The period immediately following the war and through the Khrushchev years saw Soviet strategic concerns directed primarily toward Eastern Europe and the West. Several events in the 1960s changed that. Of importance were the Sino-Soviet conflict and border clashes in the Far East, greater independence and the economic resurgence of Japan and increasing American influence and involvement in the region. As a result, the strategic importance of the Soviet gateway to the Pacific received greater attention in Moscow, resulting in substantial increases in the Soviet military presence.

This increased military presence manifested itself in a substantial build-up of troops, ships and missiles. Soviet troops in the eastern regions (including Mongolia and Central Asia) increased from 10 to 52 divisions (one-fourth of all Soviet divisions), approaching half a million troops (Swearingen, 1987, p. 250). Of these, 41 divisions were stationed in Siberia and the Far East. The Pacific Fleet, which was practically non-existent until the 1960s, was built up to become the largest of the 4 Soviet fleets, accounting for approximately one-third of all Soviet warships. Port facilities along the coast were expanded to handle the increases. One-fourth of the Soviet Air Force was deployed to the Far East. And, finally, 170 SS-20 missile launchers were deployed. These complemented the Inter-Continental Ballistic Missiles and Mid-Range Ballistic Missiles already in the region (Swearingen, 1987, p. 251).

The expansion of military preparedness in the region required investments from the military budget to handle the increased presence. These investments benefit economic as well as strategic functions. Most notable were the port facilities, especially Nakhodka, Vostochnyy, Vladivostok, Sovetskaya-Gavan', Anadyr and Petropavlovsk-Kamchatskiy, all of which, in addition to being naval ports, are also import commercial ports. Expanded facilities not only enabled the Far East to support a greatly-increased naval presence, but also to expand its commercial capabilities.

Clearly, Soviet policy over the last thirty years has been intended to secure its claims on these eastern territories, not only as part of the Soviet state but also as a region of access to the Pacific. A message has been sent that the Soviet Union is, in fact, a Pacific power. This is not to suggest that all Pacific issues have been settled simply by show of military force. Disputes still persist with China over the Amur region and with Japan over the Kurile islands. It does suggest, however, that the Soviet state is willing to protect its interests in the region and to secure Siberia and the Far East as a periphery of the Soviet Union. Recent movements toward diplomatic solutions support a changing, although not yet generally accepted, attitude toward security in the region, as in the case of a recent plea by scholars at the Foreign Ministry Academy for the Soviet Union to change its policy on the Kurile islands (*CDSP*, 27 February 1991, p.27):

> Regional security must be assured not through the special closed status and special arrangements that are characteristic of the Far East today, but through the development of independent market ties in the region and, in the long run, through integration into the regional economic structure.

Basic structure for development: the settlement system

The highly-centralized decision-making process,

and the national economic policies it has set, has produced a highly-urbanized, highly-concentrated, resource-oriented pattern of human activities. This pattern, as we have seen, is based primarily on resource extraction and transportation – river and ocean ports and along rail lines. Nowhere is this more apparent than in the settlement system, which continues to serve as the lifeline for Siberian development (Map 6.3).

By 1985, Siberia and the Far East were the most highly-urbanized regions of the country, with over 70 per cent of the population living in cities or urban-type settlements. Of the 207 cities of the Siberian and Far Eastern settlement system, Novosibirsk and Omsk had populations in excess of one million; seven more – Barnaul, Kemerovo, Novokuznetsk, Krasnoyarsk, Irkutsk, Khabarovsk and Vladivostok – had populations over half a million; and 29 cities had populations between 100,000 and 500,000. As one would expect, the functional basis of cities in the settlement system was primarily resource extraction or transportation. Several, had developed into multifunctional, administrative centres. These included Novosibirsk, Omsk, Barnaul, Kemerovo, Tomsk and Tyumen' in West Siberia; Krasnoyarsk, Irkutsk, Chita and Ulan-Ude in East Siberia; and Vladivostok, Khabarovsk, Petropavlovsk-Kamchatskiy, Blagoveshchensk and Yakutsk in the Far East. In addition to resource extraction and/or transport functions, these cities were all capitals of their respective ASSR, kray or oblast, and all had developed some level of industry.

The characteristics of urbanization in Siberia are in keeping with trends for Soviet urbanization in general: increasing concentration of urban populations in large and very large cities and the evolution of urban agglomerations. These characteristics are also in keeping with the overall nature of Siberian development: limited and highly

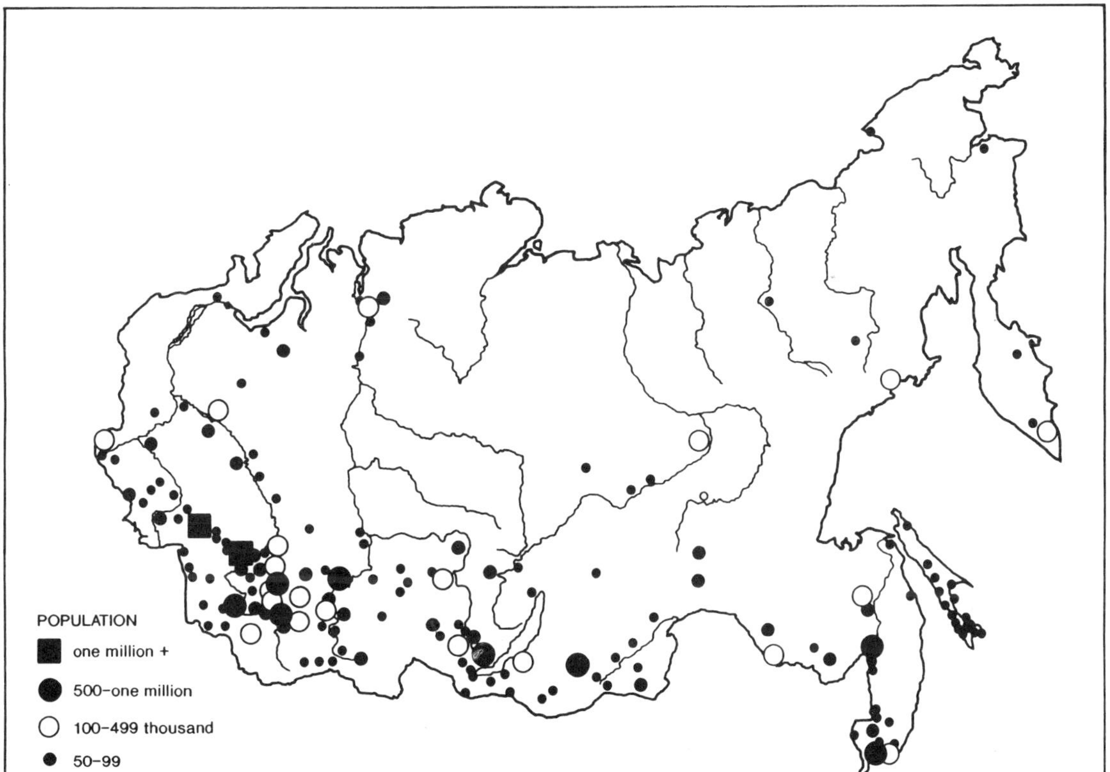

Map 6.3 *Siberian settlement system*

concentrated. The result is a settlement system that is geographically concentrated, closely related to resource extraction and transportation links, and dominated by a relatively small number of large urban centres, many of which have evolved into urban agglomerations.

According to various definitions, there are more than 25 urban agglomerations throughout Siberia and the Far East, which indicates an even greater degree of concentration than revealed by city-based data (e.g. see Pertsik, 1980, pp. 119–45). For example, Irkutsk had a population of 597,000 in 1985. Yet within its functional area, generally defined as a 120-kilometre radius of the city, there are 13 other cities, two of which have populations in excess of 100,000 each, and 5 urban-type settlements. Although data do not exist for cities with populations below 50,000, it would not be a stretch of the imagination to guess that the Irkutsk urban agglomeration exceeds one million people. In keeping with trends for Soviet urbanization in general, and in keeping with policies of limited and concentrated development, the 208 official cities of Siberia and the Far East (*SSSR*, 1987, pp. 15–17 and 23–268) are not evenly distributed. In fact, of the 208, 92 are located with the 25-plus agglomerations, accounting for 44 per cent of all Siberian and Far Eastern cities. Thus the settlement system is not only a means of implementing central government policies; it is also a reflection of how these policies have been implemented.

The highly concentrated, localized nature of development has resulted in a lack of government investment in infrastructure, which has produced sub-standard living conditions. This has made it difficult to retain labour in the region. There have been various schemes to entice labourers, with varying degrees of success. Often these schemes offer benefits to be obtained back in the western regions of the country, such as access to an apartment in Moscow. Soviet sociologists themselves have pointed out that the factor considered most important by Siberians deciding to return to the west is inadequate housing (Zaslavskaya et al., 1989).

This is not to say, however, that many are not staying; many are. By the time of Gorbachev's ascension to power, there were nearly 28 million people in Siberia, of which over nearly 20 million (71 per cent) lived in Siberian cities (*Nar khoz*, 1984, pp. 14–16). In addition to building an increasingly large urban society in Siberia, these trends have also helped to solidify Russian control of the region. From the beginning there was a great deal of cultural assimilation between indigenous peoples and the Russian overlords. This has continued to the present. Today, ethnic Russians dominate the population of Siberia, especially in the cities, and ethnic Russians alone comprise nearly 85 per cent of the population (*Ekonomicheskaya*, 1989, pp. 115, 127 and 136).

Siberia as periphery

These demographic data do not suggest that Siberia and the Far East are fully integrated economic regions of the Soviet state. On the contrary, the Russian Revolution should not be seen as an absolute break with the past. There has been as much, if not more, continuity as there has been change. The basic nature of the relationship between the Russian core and the Siberian periphery, despite the ideological rhetoric, has remained the same. What has changed, however, is the degree to which Siberia can still be referred to as a 'frontier'.

Clearly, there are regions within Siberia that can no longer be called frontiers; yet the basic nature of the relationship defines Siberia as a colony of European Russia. In this regard, the term periphery seems more applicable, because it denotes a political-economic process whereby Siberia experiences relatively 'low incomes, primitive technology, and undiversified production'. Thus, if it is misleading to refer to Siberia as a resource frontier, it may be more appropriate to refer to it as a resource periphery.

It is in this regard that we must balance local and regional impacts with national and global impacts on Siberian development. It is short-sighted to equate global effects solely to the proportion of foreign investment in the region, or to the proportion of Siberian production destined for foreign markets. Rather, it is necessary to realize that much of the domestic demand for Siberian products results from Soviet integration in the world economy: what appears to be domestic demands for development, in part or whole, may, in fact, be global. This is why the potential impacts of perestroyka present such an interesting set of possible scenarios for the course of Siberian development.

Economic restructuring and resource frontier development

Any discussion of the political economy of the Soviet Union today must examine the potential impact of economic restructuring. Although many of the particulars have not yet been spelled out, 'elements of the reforms can be pieced together' (see Gill, 1987, pp. 19–37). Economic restructuring à la Gorbachev seeks to do nothing less than radically reform the entire Soviet economy. As embodied in the directives of the 12th FYP (1986–90) and numerous pieces of legislation (see Schroeder, 1989), perestroyka might be referred to as the 'third great economic reform in all Soviet history' (Schroeder, 1989, p. 305). Only time will tell if this characterization will hold. The conservative reaction and the apparent retrenchment by Gorbachev indicate that this may become an evolution rather than a revolution. Yet change is underway and there is evidence that it will be impossible to bring this change to a complete halt.

Whatever the pace of change, several of perestroyka's components have implications for the future course of Siberian development ('Guidelines'). The key for economic development in the immediate future is modernization of existing facilities as opposed to construction of new ones. Of greatest importance will be improvements in the machine-building sector of the economy. Decentralization of decision-making and greater local autonomy is another change in direction. Imports are identified as a tool for stimulating modernization and economic growth, and special emphasis is given to the upgrading of the fuel and energy complex and agricultural sectors of the economy. This and much more is all to be accomplished within the framework of a 'socialist regulated market economy' (Schroeder, 1989, p. 316).

For Siberia, the impact of perestroyka, which emphasizes intensification and efficiency, promises to have spatial ramifications (Bond, 1987). Economic restructuring does not necessarily represent a 'turning away' from Siberia and the Far East in favour of increased investment in the European regions of the country, as some have suggested (Bond, 1987; Shabad, 1989). Apparently, grandiose schemes for development, in the tradition of the mega-projects of the 1970s, will be missing. For the time being, further development of TPC's along the BAM and water diversion schemes, for example, have been put on hold. It is not clear, however, that perestroyka will change the basic nature of Siberia's role in the national economy as the Soviet Union approaches the twenty-first century.

The first question that needs to be answered is whether the policies and changes embodied in perestroyka are going to be implemented; and if so, when, and in what form. The entire discussion of the impacts of restructuring on resource development in Siberia is predicated on the assumption that Gorbachev will succeed and that he will be able to reform dramatically the entire Soviet economy. More likely, changes in the economy will be the result of compromise – major changes, but not to the extent currently envisioned by Gorbachev and the architects of perestroyka. Even the architects themselves are not quite sure where this is leading:

> So now we have something new: the perestroyka of 1985. Why should we think that this one will be irreversible, that the changes will be long-term, that this reform will not share the fate of all the preceding ones? This is a very serious question: we might even say that it is the absolute crux of the matter . . . It is the question we must answer (Aganbegyan, 1989, p. 110).

Yet, given the momentum for change produced by Gorbachev's first five years in power, some kind of change is inevitable. The Soviet economy is embarking on a path toward reform. For some, change is proceeding too slowly; for others, too rapidly. Whatever changes do occur they will affect Siberia and the Far East, producing both continuity and change in the basic nature of the relationship between the periphery and the European core.

Siberia's role in perestroyka

> Our short and long-term plans are linked, to a considerable degree, with the tapping of the natural wealth of Siberia and the Soviet Far East ('Strategic course', p. 107).

In a series of speeches in 1986, one to the 27th Party Congress ('Strategic course', 1987), and two while on a visit to the Far East ('Text', 1987, and 'Major overhaul', 1987), Gorbachev made it clear that Siberia's role in the Soviet economy would continue to be primarily the role of a resource periphery to the European Russian core. In this respect, its role continues to be three-fold. First, Siberia will be a supplier of certain resources to domestic markets to help stimulate the economic development of western regions of the country. Second, it will be a supplier of resources to foreign markets to earn hard currency, gain technology and promote the integration of the Soviet Union into the world economy. Third, and closely related to the latter, Siberia and the Far East will continue to provide the Soviet state with access to Pacific Asia.

The primary role of Siberia and the Far East is to provide the national economy with natural resources. In this role, it is interesting to note that the recent *Ekonomicheskaya geografiya SSSR* provides a regionalization scheme based first and foremost on the contribution of the region to the 'all-union division of labor'. For domestic markets, West Siberia will continue to be a major supplier of oil and gas, coal and iron ore. Modernizing the 'fuel and energy complex' is a major goal of the 12th FYP; Urengoy oil and gas and Kuzbas coal are specifically mentioned ('Guidelines', pp. 170, 177). East Siberia will continue to be a major supplier of metals, primarily nickel, cobalt, lead, zinc and copper, in addition to aluminum and coal. The Far East will continue to provide domestic markets with fish, timber and precious metals, especially gold, silver and diamonds.

In addition to supplying domestic markets, Siberian resources will help the national economy accrue the hard currency needed to trade in the world economy. In the immediate future, integration in the world economy becomes even more important under perestroyka. Former Foreign Minister Shevardnadze made it clear that integration into the world economy is a necessary component of restructuring:

> 'In reality,' he [Shevardnadze] has said, 'the division into socialist and capitalist systems of economics, with all the extremely perceptible limitations for us, in no way signifies the absence of mutual penetration. We were drawn into the world economic process long ago . . . We should be part of the world economic system and we can and are obliged to become so . . . ' (as quoted in Aganbegyan and Timofeyev, 1988).

This attitude is in keeping with the 12th FYP Guidelines, wherein the Soviet 'import policy is called upon to actually help accelerate scientific and technological progress' ('Guidelines', p. 192), i.e. to help obtain manufactured goods, technology and hard currency. In exchange, the Soviet Union has natural resources to offer hard-currency countries. A large proportion of these resources are produced in Siberia and the Far East. Several of these resources already dominate Soviet exports abroad. Currently, petroleum and petroleum products account for 29.4 per cent of all Soviet exports; natural gas for 8.8 per cent; ferrous metals for 4.5 per cent; and timber products for 3.6 per cent. These four categories alone comprise almost half of all exports from the Soviet Union (*Vneshniye*, 1989, pp. 20-31). These are the very categories of resources Siberia supplies domestic markets. It appears, then, that Siberia will also continue to play a role in Soviet attempts to use foreign trade as a stimulus for perestroyka.

Not only is Siberia a region that will provide exports for trade: it also provides direct access to Pacific Asia, a region that may become a major supplier of imports to assist 'scientific and technological progress'.

> Given the dynamics of the international economy, the Soviet Union's position in relation to the countries of the Asia-Pacific region may be the most promising source of stimulus for change (Matuszewski, 1989, p. 6).

The economic potential of relations between Siberia and the countries of the Pacific Rim, which has been an important consideration in Russian and Soviet foreign policy for centuries, could take on even greater significance as a result of perestroyka. Gorbachev himself has clearly indicated the importance of the Asia-Pacific region to his proposals for economic change. The real significance of his 1986 Vladivostok speech, some have suggested, concerned:

> his expressed fears about maintaining Soviet economic competitiveness, his perception of the commercial and technological dynamism of the Asia-Pacific region, and his clearly articulated intention to expand Soviet trade and access to advanced technology in order to stimulate

innovation in the sluggish Soviet economy (Matuszewski, 1989, p. 10).

Yet increased Soviet involvement carries with it strategic as well as economic ramifications: the two are importantly interrelated. The Soviets have identified that world peace is necessary in order to accomplish the goals of perestroyka. This is particularly applicable to the Pacific realm, where 'the Soviet Union's identity as an Asian-Pacific power is critical to Gorbachev's twin concerns of security of the USSR and the rejuvenation of its economy' (Thakur and Thayer, 1987, p. 2).

Soviet policies directed at Pacific Asia are threefold (Samoteikin, 1987, p. 12). First, Gorbachev has called for increased bilateral relations with *all* countries of the Pacific Basin. Second, he identifies the need to settle regional disputes to ensure peace and security, specifically conflicts in Afghanistan, Korea and Kampuchea. And he proposes to scale down Soviet military presence in the Far East. The implementation of these policies would allow the Soviet leadership to secure its eastern frontier, to free up resources from the military and to gain access to the capital and technology of the region, all of which would greatly facilitate economic growth and development.

In this regard, Siberia stands ready to continue in its role as provider of resources for international trade, particularly with the countries of the Asia-Pacific region and as the bridge between the Soviet Union and East Asia. Yet within this context of continuity we may be able to discern possible changes in the nature of the relationship.

Potential impact of perestroyka on Siberia

It is not difficult to figure out Siberia's role, as perceived and planned from the core, in the Soviet economy into the twenty-first century. It is far more difficult to predict, in return, what kinds of effects perestroyka will have on the development of Siberia. To a great extent, uncertainty about the future course of Siberian development results from the fact that no one knows which policies of the overall program will actually be implemented. As a result, we can posit alternative scenarios.

The most obvious, and most cynical, prediction is that perestroyka will have little effect on the basic nature of Siberia's relationship with the core, although we shall need to wait to see whether we are talking about an All-Union or a RSFSR core. Certainly, recent policy statements and guidelines suggest that the core perceives the relationship in the same way it has for decades. The actual contributions of the region may change, and certain sectors of the Siberian economy and particular areas within the region may benefit. But these sectors and areas are those that have historically received most attention from Europe. Development will, as a result, continue to be limited, concentrated and highly dependent on the core, based primarily on the exploitation of natural resources and transportation linkages.

If some, or all, of Gorbachev's policies are eventually implemented, the scenario could be quite different: not that the perception of Siberia's role will change. Rather, several of the policies embodied in perestroyka may help to stimulate a more comprehensive development. We can only imagine the effects of greatly increasing local autonomy in economic decision-making, or allowing true competition between enterprises, based on a relatively free-market system, or individual or cooperative leasing of land. In addition to their impact on domestic affairs, these kinds of innovation in the Soviet economy could also act as stimuli to increasing the level of joint ventures with foreign corporations through free economic zones for foreign trade. Under some or all of these conditions, Siberia's development takes a different course. We might actually envision a situation wherein the rhetoric stays the same, but the reality changes.

Regionally, the future impacts of perestroyka would first affect the Far East and West Siberia, primarily because they have the best-developed infrastructure and the necessary kinds of resources, both natural and locational. Here again, these would probably be in the southern reaches of the regions. East Siberia would lag behind, but not be forgotten.

The kinds of goods exploited would remain the same; the perception of Siberia's role, from the point of view of the core, would remain the same. What would change would be the economic effects of their exploitation. In other words, under a situation of competition and free-market trading, a much greater proportion of the surplus value would remain locally. Basic industries would actually be used to build a surplus in the region, which could then be used to invest in infrastructure and other kinds of consumer-oriented activities. Three levels of economic interaction can be identified (Gill, 1987, p. 39).

At the national/international level, several products are attractive to domestic and foreign markets; these would be the basic industries, designed to produce wealth. Of primary importance are the fuel and energy resources, especially petroleum and natural gas from West Siberia and coal from East Siberia. Furs still play a role, as do precious metals. Ferrous and non-ferrous metals could be used to stimulate the economies of East Siberia and the northern Far East. And the Soviet Far East will continue to be the major exporter of timber, fish and fish products.

In addition to stimulating the development and modernization of these industries themselves, income derived from the sale of these resources could be used to stimulate the growth of other sectors of the economy, those that are primarily regional and local. These are the kinds of activities that must be upgraded if Siberia is to develop comprehensively in the twenty-first century. They include machine building, the construction industry, light industry, the consumer industry and agriculture. To become self-sufficient in some of these sectors, and less reliant in others would truly help stimulate economic development in the regions. Siberia would be using its traditional role as a treasure trove of resources and an area of access to the Pacific to finance its new role as a comprehensively-developed economic region.

Conclusion: Siberia's changing role in the Soviet Economy

These are uncertain times in the Soviet Union. The real and proposed changes of Mikhail Gorbachev foretell changes in the future course of Siberian development and the nature of the relationship between Siberia and the European core. In the near future, Siberia's role within the Soviet national economy will remain the same as it has been for this century and for over two centuries before it. The potential, given the region's natural wealth, relative location, and Gorbachev's proposed initiatives for change, is unlimited. Yet challenges from conservatives within the government, as well as demands for greater independence from the Russian republic, call into question how and when this potential will be realized. Major barriers remain to be confronted before an informed answer can be given.

We have already noted several fundamental barriers to comprehensive development. Actual implementation of the 'radical reforms' so necessary as a precondition for development is the first. The eventual resolution of All-Union and republican political and economic interests in the region is another. Traditional barriers to development have also been noted: resistance of pro-European policy-makers to invest in eastern development, lack of economic and social infrastructure, and reluctance of foreign countries toward investment in Siberia. To these we must also take into account financing and ecological concerns, both of which have been and may be affected by restructuring. Although there is not sufficient time to go into detail here, both of these are worthy of brief mention.

The program for restructuring the Soviet economy has brought with it policies of greater openness (glasnost) and local autonomy (*demokratizatsiya*), which to date have had only minor impacts on Siberian development but could greatly affect it in the future. Currently, the bulk of investment in Siberia comes from central government ministries, which concentrate on investments in industrial production at the expense of social infrastructure. In return, a great proportion of the profits from local enterprises is returned to the central ministries. Although data on this process are sparse, Schiffer provides an example whereby 98 per cent of the profits from one East Siberian enterprise were returned to the central ministry for redistribution (Schiffer, 1989, pp. 94–5). Although this is only one case, there is no reason to assume it is unusual. Changing this system and giving greater autonomy to local enterprises would enable them to keep and allocate a greater share of their profits, thus benefiting the local and regional economies to a much greater extent than presently possible.

Glasnost has also provided a forum for those concerned with the natural environment of Siberia. Although there have been celebrated cases of environmental successes in protecting the ecology of the region, e.g., Lake Baykal and the water diversion schemes, Soviet development east of the Urals has an abysmal record of disregard for the environment. An increasingly vociferous group concerned with ecological issues promises to challenge development projects in the delicate ecosystem known as Siberia and the Far East. Potentially, this concern may slow and prevent projects, greatly increasing the cost of development. To the present, however, economic development still tends to receive priority consideration over local and ecological concerns ('News Notes', *Soviet Geography,* March 1990, pp. 224–7).

So Siberia must still wait for its role to change, for the basic nature of its development to become comprehensive, designed to benefit regional and local development at the expense of national and international markets. There has been some movement toward loosening the grip of centralized control; to date, the overall impact of these trends has been minimal. For now, as it has been for the entire Soviet period, Siberia remains dependent on the core for its growth, serving as a resource frontier for European Russia and the world.

Note

2. For histories of Siberia see *Aziatskaya Rossiya,* 1974; Okladnikov and Shunkov, 1968; Semyonov, 1963.

Suggested reading

Dienes. L. (1987), *Soviet Asia,* Westview, Boulder.

de Souza, P. (1989), *Territorial Production Complexes in the Soviet Union, with special focus on Siberia,* University of Gothenburg, Sweden.

Rodgers, A. (ed.) (1990), *The Soviet Far East: Geographical Perspectives on Development,* Routledge, London.

Schiffer, J. (1989), *Soviet Regional Economic Policy: the east-west debate over Pacific Siberian development,* Macmillan, London.

Swearingen, R. (ed.) (1987), *Siberia and the Soviet Far East: strategic dimensions in multinational perspective,* Hoover Institution, Stanford.

Wood, A. and French, R.A. (eds). (1989), *The Development of Siberia,* St. Martin's Press, New York.

References

Aganbegyan, A. (1989), *Inside Perestroyka: the future of the Soviet economy,* Trans. Helen Szamuely, Harper & Row, New York.

Aganbegyan, A. and Timofeyev, T. (1988), *The New Stage of Perestroyka,* Institute for East-West Security Studies, New York.

Alekseyev, V.V. and Isupov, V.A. (1986), *Naseleniye Sibiri v gody Velikoy Otchestvennoy voyny,* Novosibirsk.

Aziatskaya Rossiya, Vol. 2 (1974) (reprint; originally published in St. Petersburg, 1914), Oriental Research Partners, Cambridge.

Bond, A. (1987), 'Spatial Dimensions of Gorbachev's Economic Strategy', *Soviet Geography,* 28, September, pp. 490–523.

Bradshaw, M. (1988), 'Soviet Asian-Pacific Trade and the Regional Development of the Soviet Far East', *Soviet Geography,* 29, April, pp. 367–93.

CIA (1985) *USSR Energy Atlas,* GPO, Washington, D.C.

Current Digest of the Soviet Press (CDSP), various issues.

de Souza, P., (1989), *Territorial Production Complexes in the Soviet Union, with special focus on Siberia,* University of Gothenburg, Sweden.

Dokuchayev, G.A. and Kozybayev, M.K. (1987), *Osushchestvleniye leninskikh idey razvitiya proizvoditel'nykh sil,* Alma-Ata.

Ekonomicheskaya geografiya SSSR, chast' II, 1989, Moscow.

Forsyth, J. (1989), 'The indigenous peoples of Siberia in the twentieth century', in *The Development of Siberia,* Wood, A. and French, R.A. (eds), St. Martin's Press, New York, pp. 72–95.

Gill, G. (1987), 'Power, authority and Gorbachev's policy agenda', in *The Soviet Union as an Asian Pacific Power,* Thakur, R. and Thayer, C. (eds), Westview, Boulder, pp. 19–37.

'Guidelines for the economic and social development of the USSR for 1986–1990 and for the period ending in 2000', report of Prime Minister Nikolay Ryzhkov to the 27th Congress of the CPSU, 3 March 1986, in Mehta, V. (1987), *Soviet Economy,* Sterling, New York, Appendix C, pp. 146–200.

Hutchings, R. 1982, *Soviet Economic Development,* (2nd edn), Oxford: Basil Blackwell.

Isupov, V.A. (1987), 'Dinamika chislennosti gorodskogo naseleniya Sibiri v period stroitel'stva sotsializma', in *Urbanizatsiya Sovetskoy Sibiri,* Alekseev, V.V. (ed.), Novosibirsk, pp. 28–45.

Itogi vsesoyuznoy perepisi naseleniya 1959 goda, (1963), Moscow.

Lho, K. (1989), 'Seoul-Moscow relations: looking to the 1990s', *Asian Survey,* 29, December, pp. 1153–66.

Lyashchenko, P.I. (1949), *History of the Nationl*

Economy of Russia, Trans. L.M. Herman, Macmillan, New York.

'Major overhaul a must – it concerns each and everyone' (text of speech delivered by Mikhail Gorbachev in Khabarovsk, 31 July 1986), in Mehta, V. (1987), *Soviet Economy,* Sterling, New York, Appendix B, pp. 124–45.

Matuszewski, D. (1989), 'Soviet reforms and the Asia-Pacific challenge', in *The Soviet Union and the Asia-Pacific Region,* Thambipillai, P. and Matuszewski, D. (eds), Praeger, New York, pp. 1–12.

Molodenkov, L.V. (1987), 'Problemy osvoeniya novykh khozyaystvennykh territoriy v Sibiri i na Dal'nem Vostoke,' in *Osvoeniye novykh khozyaystvennykh territoriy v Vostochnykh rayonakh RSFSR,* Novosibrisk, pp. 5–10.

Narodnoye khozyaystvo SSSR v 1984 g, (Nar khoz) 1985, Moscow.

'News Notes', *Soviet Geography,* various issues.

Ogawa, K. (1987), 'Economic relations with Japan', in *Siberia and the Soviet Far East: strategic dimensions in multinational perspective,* Swearingen, R. (ed.), Hoover Institution, Stanford, pp. 158–78.

Okladnikov, A.P. and Shunkov, V.I. (eds.) (1968) *Istoriya Sibiri,* 5 vols., Leningrad.

Pertsik, E.N. (1980), *Gorod v Sibiri,* Moscow.

Samoteikin, E. (1987), 'The goals of Vladivostok', in *The Soviet Union as an Asian Pacific Power,* Thakur, R. and Thayer, C. (eds), Westview, Boulder, pp. 11–18.

Schiffer, J. (1989), *Soviet Regional Economic Policy: the east-west debate over Pacific Siberian development,* Macmillan, London.

Schroeder, G. (1989), 'Soviet economic reform: from resurgence to retrenchment?' *The Russian Review,* 48, pp. 305–19.

Semyonov, Y.N. (1963), *Siberia: its conquest and development,* Trans. J.R. Foster, Hollis & Carter, London.

Seton-Watson, H. (1967), *The Russian Empire, 1801–1917,* Clarendon, Oxford.

Shabad, T. (1989), 'The Gorbachev Economic Policy: is the USSR turning away from Siberian development?', in *The Development of Siberia,* Wood, A. and French, R.A. (eds), St. Martin's Press, New York, pp. 256–60.

Shabad, T and Mote, V. (1977), *Gateway to Siberian Resources (The BAM),* Wiley, New York.

SSSR Administrativno-territorial'noye deleniye soyuznykh respublik, 1987, Moscow.

'The strategic course: acceleration of the country's socio-economic development' (excerpts from the political report of the CPSU CC to the 27th Congress of the CPSU, presented by Mikhail Gorbachev), in Mehta V. (1987), *Soviet Economy,* Sterling, New York, Appendix A, pp. 86–123.

Swearingen, Rodger (1987), 'The Soviet Far East, East Asia and the Pacific', in *Siberia and the Soviet Far East: strategic dimensions in multinational perspective,* Swearingen, Rodger (ed.), Stanford: Hoover Institution, pp. 226–72.

'Text of speech by Mikhail Gorbachev in Vladivostok, 28 July 1986', in *The Soviet Union as an Asian Pacific Power,* Thakur, R. and Thayer, C. (eds), Westview, Boulder, Appendix, pp. 201–27.

Thakur, Ramesh and Thayer, Carlyle, (eds.) (1987), *The Soviet Union as an Asian Pacific Power: implications of Gorbachev's 1986 Vladivostok initiative,* Westview, Boulder.

USSR Energy Atlas (1985), CIA, Washington, D.C.

Vneshniye ekonomicheskiye svyazi SSSR v 1988 g (1989), Finansy i statistika, Moscow.

Zaslavskaya, T.I., Kalmyk, V.A., and Khakhulina, L.A. (1989), 'Social development of Siberia: problems and possible solutions', in *The Development of Siberia,* Wood, A. and French, R.A. (eds) New York, St.Martin's Press, pp. 177–87.

7

Environmental degradation in the Soviet forest

Brenton M. Barr

Introduction

The Soviet economy's performance in allocating scarce resources is unable to overcome major shortfalls in the country's physical and ideological requirements for the products of a modern industrial economy, particularly one associated with a military superpower comprising one sixth of the earth's land surface and a major portion of its industrial raw materials, including one quarter of its standing timber reserves. More than seventy years of activity have also created a spatial economy in which links between places of supply and demand have been forged on the basis of administrative and managerial expediency rather than economic effciency and reliable planning criteria; hence numerous regional economic activities and functions are incorrectly and inappropriately located over space with the result that the costs of their operations divert capital required for incremental output of industrial production. On an expanded scale, however, particularly in the realms of environmental protection and viability of human settlement, spatial characteristics of the Soviet economy are adversely affecting broad regions of the country's physical and human ecology, perhaps even the ability of vast areas to support human life itself. Nevertheless, expenditures on forest conservation or protection are declining as a per cent of total Soviet Union outlays for environmental protection. In 1985, 1.04 or 11.4 per cent of the total output of 9.13 billion roubles were expended on the forests (Goskomstat 1989, p.9); in 1988 1.12 or 10.1 per cent of the total 11.1 billion roubles went to the forests.

Since the rise to power of Gorbachev in March 1985, the Soviet Union and the world have been subjected to extensive exposés of the need for economic, political and environmental reform in the Soviet Union. Implicit in these domains is the need for vastly enhanced management of the spatial economy and the environment which by definition has a spatial character. For the first five years of his rule, Gorbachev in essence set the rules for openness and debate, although it was clear to many from the outset that he lacked a coherent blueprint for the future structure of the Soviet Union or an effective algorithm for its creation. By the fifth anniversary of Gorbachev's assumption of power, other leaders had arisen within parts of the Soviet Union whose demands for participation in setting parts of the national agenda for change could not be ignored. By early 1989, therefore, one could refer to the 'first post-Gorbachev era' in which the Soviet Union's president was required to share determination of the agenda for change with strong regional leaders like Yeltsin, Polozkov and Landsbergis, and in which long-unresolved ethnic and territorial issues prevented central authorities from exercising customary autocratic power. Regional dissent and dissenters are now found in all parts of the Soviet Union, including the RSFSR; in aggregate they reveal that the economic, political, cultural and ethnic problems of tsarist Russia in many cases have been in limbo for over seventy years despite the much-vaunted military and industrial successes of Lenin, Stalin and their successors. In short, too many unresolved issues with a strong geographical underpinning that have been ignored, repressed but especially unresolved since 1917 are now coming to the fore with unbelievable speed and are often diverting all leaders' attention from a host of truly pressing economic priorities. In many cases actions and programs associated with the industrialization of Russia and the Soviet Union may have exacerbated pre-existing problems, but in all cases it is clear that they have created a legacy of dysfunction which is many orders of magnitude greater in complexity than that which existed in 1914. In sum, the Russian empire's problems and geographical dilemmas on the eve of the First World War appeared capable of solution; the challenges facing the Soviet empire on the eve of the twenty-first century in many cases may defy resolution although in both epochs opinions on remedies are as diverse as the participants themselves. At present, one glimmer of hope common to many of the Soviet Union's crises is that solutions may be found through regional autonomy and greater regional and local control and decision-making, in contrast to the panacea sought by Stalin and his heirs in rigid central political and economic control.

Reform of the Soviet political and economic system is thus needed because performance is now recognized as being totally inadequate for the leaders' and citizens' expectations and basic requirements. But from a geographical and environmental perspective, why is reform needed?

Although numerous environmental disasters and planning failures, such as Chernobyl and the utilization of water in Central Asia, are well documented in the literature, and although myriad shortcomings in the Soviet system have been revealed through glasnost (openness) as needing perestroyka (basic restructuring), the majority of analyses seem to approach the current changes and dilemmas in the Soviet Union from an economic, historical, political, sociological or ethnic point of view. The geographer, on the other hand, sees the great physical and human diversity of the Soviet Union as underlying all topics of current concern and as having a symbiotic relationship with the myriad patterns and processes of interaction, or previously proscribed spatial linkages, comprising the spatial economy and its attendant transportation, industrial and settlement systems. The Soviet experience is highly spatial or geographical because of the vast distances separating its constituent parts and the great inherent diversity in resources, environments and peoples among its regions. The country's forest is largely found in the Russian republic (Map 7.1), but from Karelia to Kamchatka the growing stock varies immensely in age, stocking density, species composition, site quality and accessibility. What may be commercial or industrial timber along the major rivers and railways in the European North and Northwest or western Urals, for example, may be totally unusable in Siberia and the Far East due to the adverse impact on its delivered value imposed by transportation, the greater cost of harvesting and local movement, the prevalence of species with limited market potential, the adverse environmental conditions affecting regeneration and the rotation cut, or the dominance in much of the east of mountain forests more reminiscent of western North America than of the traditional environment in which almost all Soviet, particularly Russian, forest practices have evolved. Even if the Soviet forest were uniform in all natural respects from border to border in the physical properties of growing stock and environment, some regions would be much more highly valuable than others because of the adverse cost of the friction of distance associated with movement of roundwood and products to market. In a Soviet planning

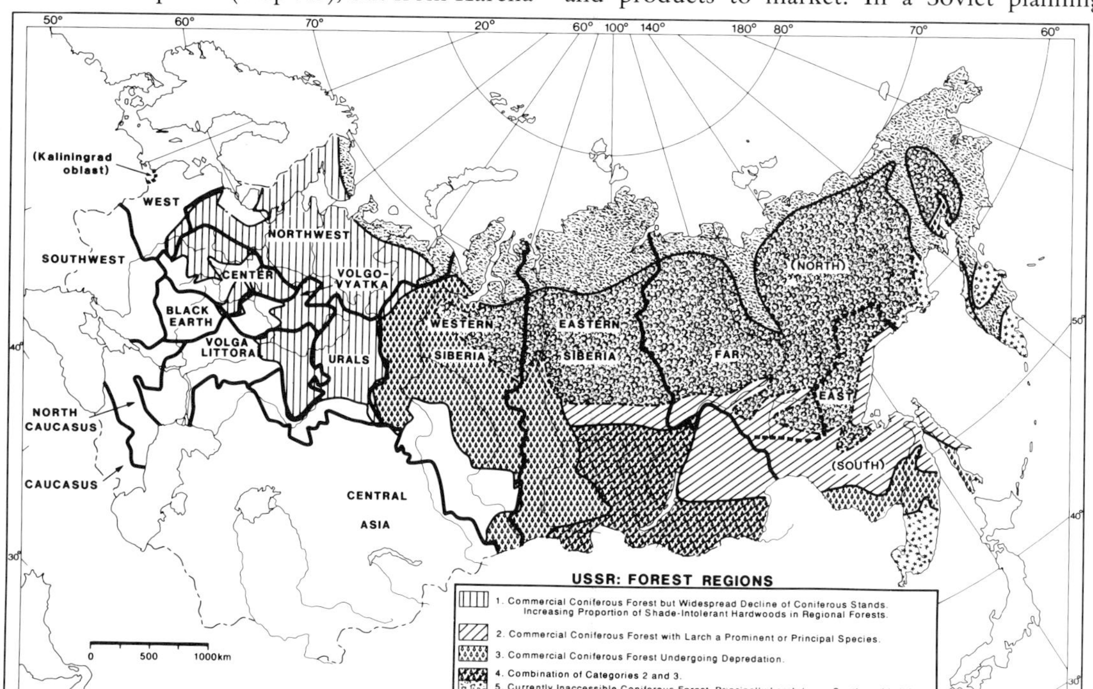

Map 7.1 *USSR forest regions*

context and the customary constraints of capital investment and transportation infrastructure, a uniform forest would still have great spatial variation in utility.

The quality of the Soviet growing stock and regional environments is not only heterogeneous, but its constituent parts have been treated unevenly through time. The European-Uralian forest has come under the longest and most sustained period of utilization and probable extent of overall depredation due to developmental exigencies, war and toxic emissions from other economic activities. Forests in the east, however, have been remote from society's awareness for much of the Soviet period and have been subjected to rapacious harvesting, often by numerous competing administrations in the same forest. Accessible timber in the southern Soviet Far East, likewise, has been far from probing eyes, but has also suffered from the actions of foreign intervention and other predatory activities. Despite much official rhetoric to the contrary, the Soviet forest has suffered from officially sanctioned overcutting leading to depletion of growing stock and imbalance among stands and species, and from expediency on all levels engendered by a ruthless development-at-any-cost ethos and adversarial military relations with the Western industrial world. The Soviet practice of centralized administration has not been synonymous with the goals of central planning, or indeed of planning *per se*. Planning agencies have been facilitating the goals of central ministers and politicians to fulfil plans if possible, and at any cost if necessary. Rents have been reassigned from profitable sectors and regions to fulfil national and international military and political goals, and to ensure an above-average standard of life particularly for residents of Moscow, Leningrad and Kiev (Medvedkov, 1990). Gorbachev's perestroyka may not have started as a move to ensure regional equality within the Soviet Union or to enhance the standard of living of nearly 300 million people, but its unintended outcome is likely to change established spatial relations and patterns of resource utilization forever as republic, regional and local politicians started in early 1989 increasingly to determine the political agenda and determine economic priorities.

Whither reform? An agenda for analysis

Soviet and Western literature is replete with details and analyses of Gorbachev's and other leaders' attempts to resolve problems, establish priorities and effect widespread reforms in the Soviet system on all fronts. Detailed aspects of change in forestry and the forest product sector found throughout the Soviet forestry literature and reported elsewhere (Barr, 1989) supplement several decades of investigation by the author and other analysts. The daily press and myriad specialist periodicals continue to provide detailed analysis of issues and problems, including those related to forests and the environment that may be consulted for their up-to-the-minute currency (detailed analyses of the sector and its possible future are provided by United Nations 1989a and 1989b). This plethora of detail often appears, however, to be disparate, diverse and disjointed; it begs for identification of coherent themes and underpinnings and for synthesis within the wider context or broad frame of Soviet politics, economics, environment and geography. The present chapter, therefore, offers a timely opportunity to link the Soviet reform movement initiated by Gorbachev after 1985, and now increasingly dominated by other leaders and members of the Council of People's Deputies and its standing commissions, to sectoral and regional problems in the Soviet forest and to the vast legacy of decades of environmental degradation in major sectors of the Soviet economy, including the forest.

This chapter presents an opportunity to reflect on how the major concerns of Soviet reformers have ineluctable implications for the country's geography, including spatial or regional relations and the environment. Rather than focus on the analysis of specific issues, many of which are currently viewed in depth in the specialist literature, this chapter attempts to synthesize the political and economic priorities of reform and stated reconstruction in a spatio-temporal framework using forestry and the forest product sector as a foil. Probably no resource or form of economic activity is as widespread in the Soviet Union, particularly Russia, as the forest industry, and hence its problems and prospects seem to offer legitimate scope for a wide-ranging spatial synthesis that picks up on otherwise disparate threads of new thinking (*novoye myshleniye*) and all the revelations prompted by glasnost which, when initially proposed by Gorbachev were supposedly intended to lead toward perestroyka of the Soviet economy but which instead, or at least in part with the forces thus released, may have helped induce chaos and disintegration (*razval*) in a system already suffering from terrible sectoral and spatial dysfunction.

By no means exhaustive, the chapter's discussion of environmental problems complements the detailed analyses of a broad spectrum of environmental issues by scholars like Micklin and Bond (1988), Pryde (1991) and Ziegler (1987), including those identified as critical in the 1990s (Panel, 1990).

The overwhelming message of this chapter is that the Gorbachev era initiated change on a myriad of fronts that was often only partially executed, and that has left a host of continuing dilemmas and challenges for new regional leaders and Gorbachev himself in the first post-Gorbachev era. Although beyond the scope of this chapter, many of Gorbachev's administrative reforms have the appearance of 'new bottles for old wine' (Barr, 1989, p.15) because they have affected the form but not the substance of the Soviet economy and hence are not true reforms. At the moment, the Soviet economic system acts like a planned socialist economy without a plan and perhaps without socialism, and is about as effective as a capitalist economy without a market or convertible currency.

The importance of geography

If the Soviet system does not function according to aggregate expectations, its regional and environmental shortcomings are perhaps even more acute because they can no longer be masked by a host of counteracting variables. Thus crowded housing, for example, has for long been offset by social security and guaranteed employment. Although delays in obtaining items of social consumption, including automobiles, washing machines, dishwashers and video recorders are frustrating to individuals, hope is maintained among the population through waiting lists and occasional mass importation of the items or the means to produce them. This hope persists even today, although not with the same outward acquiescence as before. The consequences of many environmental malpractices, however, have long been inescapable although they were obscured in official secrecy. Pollution of Lakes Baykal and Ladoga, for example, by forest-sector enterprises should have been halted decades ago. Inadequate forest-fire suppression went unreported until 1988 because forest-fire statistics were highly classified, perhaps because of their possible links to sabotage, or because they comprise one of the most important indictments of Soviet resource and environmental administration. Extensive damage to the tundra and taiga from northern non-ferrous metallurgical activities long went unreported, although the spatial extent of forest degradation is now admitted to cover millions of hectares of coniferous forest and to affect the life sustainment and economic viability of much of Karelia and Eastern Siberia, including the bountiful forests of the Yenisey-Angara river system and hundreds of thousands of hectares surrounding the Bratsk and Ust'-Ilimsk forest product enterprises. The formerly beautiful pine forests lining the highway between the settlement nodes of Bratsk are now a ghastly brown, testimony to the rapid environmental destruction that has occurred in only two decades.

Until the 1980s, resource exhaustion or environmental degradation was masked by the shifting of extraction and harvesting to other locations. Almost in a gargantuan grotesque parody of primitive slash-and-burn activities of tropical agriculture, the Soviet Union rapaciously utilized first the forests of northern European Russia and the Urals, and then attacked the forests of accessible Western and Central Siberia and southern districts of the Far East. Although the country continues to claim large areas of growing stock, harvesting practices of the past seventy years have sapped the commercial viability of much of the country's accessible forests; measures to ensure their regeneration have been half-hearted at best. Rents from the forest and income derived from wood processing have either been insufficient to ensure sustained yield management, diverted to support other priorities of the state budget, or used to subsidize logging and processing enterprises within the forest sector itself. Even in mid-1990, nearly half the main logging operations of *Minlesprom* (Ministry of Forest Products) require subsidization to remain in operation. In addition, substitution of low-grade timber and less desirable species has been largely dismissed in favour of greater exploitation of peripheral stands despite the unrealistic costs of their utilization. By 1989 the Soviet Union was incurring an annual shortfall in the supply of roundwood of twenty five million cubic metres, or 6 per cent of its annual demand. The magnitude of this shortfall is similar in 1990; the country now admits it may undergo a shortage of paper and paperboard of 500,000 tons, or 5 per cent of its already modest requirement. With the world's largest reserve of standing timber, but a rank in per capita consumption of paper products of 36 among the world's nations, the sectoral and geographical shortcomings of the Soviet economy are probably nowhere as apparent as in forestry and forest-products (and increasingly in the petroleum sector). The forest sector thus is a clear example of why economic reform is inseparable from geography – that is, from existing regional dysfunction and cumulative environmental degradation. Prior to Gorbachev, the growing stock of peripheral regions appears to have had an unquestioned role in Soviet development, particularly in the extensive use of resources and assimilation of undeveloped regions. Gorbachev's initial reactions to eastern development, for example, were clear pronouncements that the long periods of construction and excessively large volumes of capital required to bring developments on stream should be avoided and that alternatives in other regions or abroad should be pursued. Hence volumes of timber that hitherto might have been expected to support primary manufacturing in various industrial complexes along the Baykal-Amur Mainline (BAM) railway, for example, were not accorded economic utility as their value in effect exceeded what the market could bear. On the other hand, under Gorbachev enthusiasm for joint ventures with foreign firms has risen; between January 1987 and April 1990, 97 were registered in the chemical and wood products sector out of a total of 1447 for all sectors (*Soviet Joint Ventures*, 1990, p. 14). (Customary Soviet purchases of technology, however, continue through non-joint venture technology transfer; see Chapter 9.) Many forest-sector joint ventures involve utilization of process technology developed abroad, in Finland, Scandinavia and Canada for example, suitable for efficient utilization of species and log grades previously ignored by Soviet industry. Much of this timber is region specific, such as for example deciduous species in European Russia and small diameter pine and larch in East Siberia and the Far East. By encouraging joint ventures which in effect are associated with particular regions, Gorbachev is ensuring that his modest reforms to date have an important geographical dimension with attendant consequences for regional development and the environment.

The forest as microcosm: who is in charge?

Under Gorbachev, control over economic activity and the environment has become diffused and chaotic. Various territorial administrations have been accorded the opportunity to deal directly with foreign companies and to conclude joint ventures relatively independent of the central planners. The forest-product sector dominated by *Minlesprom* was reorganized in March 1988 to decentralize industrial management and enhance coordination of utilization procedures with local authorities; *Exportles*, the foreign trade organization for forest products, was transferred to *Minlesprom* in Moscow as part of a general reform of the foreign-trade sector. This reform was also intended to enhance reforestation, although experience to date confirms early suspicions that funding of forestry activity by *Minlesprom* would be inadequate. *Minlesprom* itself in mid-1990 was largely transformed into a giant corporation whose Moscow activities might resemble those of a multi-national firm in scope and function, while many of the member territorial enterprises ('sister companies' in Western parlance) would pursue market and production opportunities relatively unfettered, including those with foreign companies. Much of the technical expertise required for broader corporate planning and analysis could be provided by *Minlesprom*, while maximization of output, utilizations of the factors of production and export opportunities would be the responsibility of member enterprises. If this proposal was realized, and if state orders (*Goszakazy*) and ministerial intervention were truly minimized, then member firms might effectively compete for capital, resource allocations and market opportunities. Timber agreements to date concluded or pending between the Soviet Union and Japan, Canada, the United States, China, Australia and South Korea, suggest that foreign-market opportunities for logs, lumber and other forest products previously unrealized by central Moscow authorities in the Asia Pacific region, for example, may now become viable. In this sense, the regional geography of that part of the forest sector with joint participation may become more market-driven and its spatial structure more dependent on resource quality, local environmental concerns and the profitability of transportation and product mix. The regional geography of that part of the sector – by far the largest now or in the foreseeable future – serving the internal Soviet market, however, will probably be driven by priorities seeking to enhance resource recovery from accessible regions, reduce the demand for new settlement and transportation facilities, and overcome the serious shortage of paper and board products by Soviet consumers and industries. Some of the pressure for changes in spatial relations within the forest sector may be motivated by the growing inability or unwillingness of the Soviet railroad system to supply the forest sector with railway cars – in excess of 160,000 per annum in recent years. The reluctance of the railway to move timber and forest products may be due in part to congestion, shortage of rolling stock and inadequate infrastructure, but may also have to do with the inherently low value of logs and lumber, and what may be the railway's inability under the new economic conditions to subsidize unprofitable hauls (see Chapter 8).

Throughout most Soviet industrial history, major investment and production priorities have been formulated by planning periods and effected by central ministries and state committees subject to approval and verification by the Communist Party of the Soviet Union (CPSU) through its periodic congresses and central committee plenums. The extent to which this established system of allocating resources is still operative is presently unclear given the preoccupation of the 13th Party Congress in mid-1990 with struggles for power, maintenance of national unity and myriad problems related to regional nationality unrest and ethnic animosity. The established Moscow-based command economy is at risk, and after expiry of the 12th FYP at the end of 1990, appears to be undergoing replacement by less formal centrally inspired production arrangements, or myriad economic links negotiated by firms within and among republics subject to the overall priorities of territorial administrations at all levels – local, regional, republic and All-Union. The Soviet system appears to lack coherence and unity. Hence the spatial utility of its constituent parts and resources, including the forest's regionally diverse growing stock, varies according to the priorities and policies of the increasingly strident and independently minded leaders of union republics. Relations in the past which were governed by central decree (*dictat*) increasingly will have to be valorized according to the needs of those who control them, and denominated in values agreeable to all (such as United States dollars and German marks). Scarcity of convertible currency, for example, could cause territorial units like Estonia or Lithuania to enhance silviculture and recovery of timber through various forms of intermediate cut rather than to purchase timber abroad from Russia, for example. On the other hand, the Russian republic's vast growing stock, although a large portion is in currently inaccessible areas, could lead that republic to continue to export roundwood to some parts of the Soviet Union in exchange for food and high-technology goods, to export roundwood and lumber to established Soviet markets in industrial countries for foreign currency, to employ joint ventures for development of new markets and to continue to use timber on a barter basis in border trade with China. At the same time, an independent Russian republic with reconstituted environmental and conservation priorities could emphasize the rehabilitation of its European-Uralian forest, enhanced utilization and recovery from both coniferous and deciduous species. It is unlikely that the geographical dimensions of forestry and the forest sector in a more sovereign Russia would differ considerably from those of the Soviet Union within the present Communist-dominated system. If, however, Russia were to assign economic and environmental values to its resources, including the forest, more in keeping with quality, location, movement, replacement and preservation, then the geography of forestry and the forest sector could drastically shift toward more intensive utilization and management of the European-Uralian forest. Such actions could also be taken by the Soviet Union, but the initiative for major economic reform seems to have slipped away from central federal authorities and to have passed for implementation instead to republic and perhaps regional and local leaders. Ownership or some type of formal responsibility for maintenance and stewardship of the forest, however, would have to be assigned to ensure that those in charge indeed had the incentive and legal authority to ensure its preservation and efficient use.

The reorganization of formal responsibility for environmental protection under Gorbachev's leadership in January 1988 with the creation of *Goskompriroda* (the USSR State Committee for Environmental Protection) is indicative of the need for a comprehensive approach to the environment and natural resources in the Soviet Union, and of the extent to which initial measures so often are enthusiastically proclaimed and then not rendered effective through completion of enabling administrative and legislative measures. Before the establishment of *Goskompriroda*, environmental protection had a relatively minor status within *Gidromet* (the USSR State Committee for Hydrometeorology and the Environment) but many of the actual environmental protection functions were fragmented among various ministries and departments, including Agriculture (the State Agro-Industrial Committee), Forestry, Fisheries, Geology, and Land Reclamation/Water Resources (Sagers, 1988, p. 635). *Goskompriroda* assumed the functions and was accorded appropriate subdivisions of numerous departments and ministries, including parts of *Gidromet. Goskompriroda* was granted extensive powers (described by Sagers, 1988, p. 636) 'to protect the environment and organize the rational use of natural resources in the USSR' and was 'together with the Ministry of Justice . . . instructed to prepare, in consultation with other ministries and departments, a draft law on environmental protection and submit it to the Council of Ministers in 1989'.

By late 1989, however, Sagers (1989, pp. 684-5) could write that 'the powers and mandate of the legislative environmental committee are particularly murky at this time . . . . *Goskompriroda* has been virtually moribund since being established.' Its head resigned in June 1989 apparently after losing a power struggle with the politically powerful head of *Gidromet*, Yuriy Izrael' (who was still powerful and restricting the full operation of *Goskompriroda* in mid-1990 according to informal remarks in Harrogate, England in July 1990 by Igor Altshuler, Moscow University geographer and co-founder and vice-president of the Association for the Support of Ecological Initiatives). The new chairman of *Goskompriroda* is not a member of the CPSU and hence may lack the political prowess necessary to ensure that environmental protection receives top priority within the Council of Ministers where immediate economic considerations, worsening with each passing day, will probably continue to be paramount.

The fate of environmental protection in forestry both before and after the sector's industrial restructuring in March 1988 is virtually identical to that of the entire country as represented by the experience since January 1988 of *Goskompriroda* (including its supposed responsibility for forestry). Fragmentation of responsibility for the environment and natural resources under *Gidromet* 'as well as the incongruous situation in which the same ministry or department charged with exploiting the resource was also supposed to protect it' were 'long cited as reasons for the relatively weak nature of Soviet environmental protection efforts' and the basis for 'calls to establish a separate, unified agency for environmental protection in the Soviet Union' (Sagers, 1988, p. 635). In creating *Goskompriroda*, the CPSU and the Council of Ministers cited '"excessive fragmentation" of activities among the various ministries and departments, which did not meet current needs and had become a "serious factor holding back the intensification of production in conditions of growing interdependence between the . . . environment and the development of the economy"' (Sagers, 1988, p. 635). They observed that the behaviour of various departments and ministries toward the environment is irresponsible and has frequently led to regional environmental degradation, often due to the 'residual principle' in allocating material and financial resources to environmental protection. Essentially, the production ministries, under pressure to fulfil output plans, have long directed available funds and resources into expanding production; environmental protection activities received whatever was left over (Sagers, 1988, p. 635).

If the terms 'reforestation' and 'forest regeneration' are substituted for 'environmental protection' in the last quotation, then the experience of forestry and the general environment are identical in the Soviet Union. Just as the formal establishment of *Goskompriroda* in 1988 has not improved the Soviet environment (particularly as environmental legislation is still absent), so the forest sector restructuring of that year has not resolved the hiatus between loggers and silviculturalists, particularly in the major forest regions of the RSFSR where *Minlesprom* is formally responsible for both activities!

The Soviet forestry and forest-product sectors were reorganized on 10 March 1988 as part of Gorbachev's perestroyka. *Minlesbumprom* was transformed into *Minlesprom,* and *Gosleskhoz* became *Goskomles*. The middle level of production amalgamations in *Minlesbumprom* was eliminated, and reforestation in major timber-harvesting areas became the responsibility of *Minlesprom* to promote the current objective of 'comprehensive timber utilization'. These and related administrative changes in the forest sector are part of the much-discussed drive toward improved economic performance, including greater value derived per unit of raw material consumed, the elimination of deficits among numerous enterprises (at least half of the country's logging operations lose money) and effective regeneration of the commercial growing stock.

Other events related to forestry since early 1988 include the formulation by *Goskomles* of a draft plan for the period through to the year 2005 and the establishment under the auspices of the Academy of Sciences' Department of Philosophy and the USSR Philosophy Society [CDSP, 40 (49), 1989, p.30] of a popular body, the State Ecological Fund. As in numerous other sectors of the Soviet Union's economy and areas of its society, the winds of change are strong, long overdue, but rather unpredictable. With or without perestroyka, however, responsible Soviet administrators are increasingly aware of environmental deterioration throughout their country and are realizing that the assault on the human environment by continuing industrialization is wreaking ineluctable damage. The major question at this juncture, therefore, relates to the extent to which change and possible restructuring will enhance or exacerbate the environment, particularly that of the Soviet Union's forest.

A view of the future: perestroyka to 2005

Major issues continue to be endemic throughout the forest sector. To what extent, therefore, has the new way of thinking permeated the managerial system and become effectively incorporated in planning through to 2005, the new target date associated with long-range planning? The following discussion examines recent materials, particularly from the forestry press, for evidence that restructuring to date, or at least movements toward restructuring, have had, or are likely to have, a salutary influence on the problems discussed above. This discussion proceeds, however, with the caveat that a large system such as the Soviet forest sector is unlikely to undergo radical change during the course of only a few years, particularly when so many attendant pieces of legislation and well-established social values among the populace remain unaltered.

Goskomles issued a draft of its idea (*kontseptsiya*) of perestroyka to the year 2005 on 5 January 1989, ten months after the forest sector was restructured. The draft is in keeping with the decree on the sector's restructuring that emphasized unified management of the timber resource, the broad ecological, economic and social significance of the country's forests, the need for unbroken continuity between timber-harvesting and regeneration and the shift toward economic rather than administrative forms of management through effective branch and territorial systems.

The *Goskomles* draft occupied a full page of *Lesnaya Promyshlennost'* (5 January 1989, p. 2) and concluded by asking the newspaper's readers to comment on the proposals and write directly to it or the committee.

New economic measures

In the future, all aspects of the forest resource will

generate income, not just the provision of timber. This will generate a greater revenue base for managing the forest and will be broader than stumpage, service charges and fines in order to offset costs to the state budget for conservation and related measures. Stumpage will continue to be important. At the moment, however, all stumpage accrues to the state budget and hence enterprises are not stimulated toward efficiency. Starting in 1990, higher stumpage fees were to provide a major addition to forest income. Forestry workers were to be allocated not less than 15–20 per cent of the allowable cut to produce mature timber which they will be able to market competitively or produce commercial products. A second major source of income is to be derived through leasing contracts with other branches of the economy, cooperatives, social organizations and individual citizens. Leasing should be extended to numerous spheres.

Economic accountability should be introduced in all enterprises based on planning-calculated prices differentiated according to the quality of achieved results. The major organizational form of labour should be the collective contract. For fulfilment of ecological measures to avert desertification in Kazakhstan and Central Asia, provision of wooded pastures and securing sediments on former parts of the Aral Sea should proceed in special projects financed from the state budget. Capital pools should be created for insurance and incentives in forestry from monies collected through fines, disciplinary measures and forfeits. Forestry has the weakest capital base (*fondovooruzhennost'*) of any branch of the Soviet Union's economy. The volume of capital investment in this branch is excessively low and is inadequate for its development. The branch requires an annual increment of 10–12 per cent to its capital assets and a speeding up of the output of forestry machinery. The branch's machine-building factories today have a low capacity and are poorly equipped with repair facilities. They need to be rebuilt.

Silviculture

Although the Soviet Union annually practices silviculture on over 2 million hectares, it loses on average 2 per cent of the area of forest plantations – 1.3 per cent in Europe and 2.6 per cent in Asia. An undesirable change of species occurs on most of these lands. Greater work is required in seed production and plant breeding; to date only elementary steps have been taken. Selection facilities and nursery greenhouses should be expanded to provide foresters with higher quality planting stock. The Soviet Union currently produces 5.5–6 billion seedlings and saplings annually which is adequate for forest plantations, but frequent shortages occur in the provision of replacement stock to linked agencies. *Goskomles* proposes to produce 8–9 million containerized seedlings annually by 2005.

Silviculture must be transformed into a self-financing system in which wages reflect achievements related to regionally defined silvicultural goals. *Goskomles* states that the Soviet Union has 315 million hectares of agricultural land requiring protection by trees, but existing plantations constitute only one quarter of the required amount. By 2005, the country needs to plant 3.9 million hectares of protective plantations, including 660 thousand hectares of shelter belts protecting fields. These all should occur in the form of *goszakazy* because appropriate land areas must be simultaneously designated for this activity. At the same time, existing protective plantations must undergo improvement and reconstruction and wooded pastures must be expanded in desert regions. Silviculture should also benefit from enhanced land drainage (hydro-amelioration), but first the existing drainage network must be repaired and restored. Some 200 thousand hectares should be drained.

Changes in the next ten years will occur in the material-monetary evaluation of allowable cut and in harvest plans. The forest reserve will undergo a cadastral evaluation (*kadastrovaya otsenka* – extent, value and ownership/control) and annual harvesting will proceed on the basis of enhanced data analysis of all major aspects of the forest. Creation of a comprehensive long-term utilization and expanded-reproduction program is envisaged. Forest management designs should constitute legal and technical documents with defined goals and long-term strategies, technical and economic standards (norms), and should be coordinated with planned forestry by forest enterprises in the remainder of this planning period and the next five-year plan in keeping with tasks already established for each oblast, kray, ASSR and SSR.

Comprehensive utilization

Forest utilization should comprise all aspects of the resource: wood, non-wood materials and all valuable elements of the forest. New products should be manufactured, such as wafer board. Even commercial hunting should become part of the comprehensive use of the forest; it now provides a return of 70 roubles per thousand hectares, or one tenth that of most European countries. A comprehensive and *centralized* approach to hunting should increase the supply of game meat by 15–20 per cent by 1995 and by 30–40 per cent by 2000 and also facilitate the preservation and enhancement of wild-animal habitat. Recreation is an important component of comprehensive forest utilization. Hence, the status of nature preserves and national parks should be enhanced and at the same time their management should be *centralized* to ensure a unified scientific-technical policy and approach to forestry in such special conservation areas.

Forest protection

Contemporary conservation levels do not safeguard future requirements. Furthermore, some critical phenomena are already appearing related to the widespread use of chemicals and zones of radioactive pollution related to the Chernobyl AES. Nevertheless, measures should be introduced for practicing forestry and using products derived from polluted areas. The area of the state forest reserve regularly subjected to stand evaluation should be expanded for assessment of forest diseases and pests. Active intervention should occur in forest pathology to eliminate sources of disease and to halve the damage caused by pests.

Forest-fire control should be practised in all forests with a coherent management structure. By 2005, forest-fire control should be exercised on 1.157 billion hectares (the state forest reserve comprises 1.259 billion hectares or 56 per cent of the Soviet Union's land area (*Lesnaya Promyshlennost'*, 17 December 1988, p. 2)) and focus first on reducing the growth in the number of fires started by human carelessness, and then between 2000 and 2005 gradually reduce the number of fires by 5–10 per cent. This will reduce overall the annual economic losses incurred through large-scale fires by 60–70 per cent.

Accelerating scientific-technical achievement

Goskomles' potential for improved performance is considerable, but to date the research findings have had little effect. Technical policies should be reformulated to improve the sector's technology, organization of forestry and timber utilization. This process should first create a program-approach to the forest which unifies scientific potential, financial, material and technical resources of forest administrations and institutions of higher education and directs them toward consecutive solution of the most pressing problems. Fundamental reconstruction of science toward raising the effectiveness of its research will facilitate transfer of the branch's institutes to economic accountability and self-financing and encourage them to compete for research contracts, including the development of direct foreign links.

Goskomles proposes to raise the effectiveness of research and develop closer ties with producers by organizing scientific production amalgamations. The study of problems of mountain forests (their area constitutes 40 per cent of all forests) will be concentrated in an All-Union mountain-forest scientific-research institute which will be developed on the basis of the Institute of Mountain Silviculture. Research stations will be opened in the vicinity of the Aral Sea, the BAM service sector and in Western Siberia. Their mandates will be carefully stated and they will comprise specialized subdivisions studying the production of growing stock, particularly Siberian stone pine, on a selection basis, and the introduction of new technology in production. The activities of the institute will be measured by its achievements.

The design of new equipment takes too long and consequently increases in labour productivity are minimal. The quality of studying design ideas is low and an absence of incentive causes scholars not to seek major accomplishments. Consequently, all scientific-research institutes associated with aspects of mechanization are to begin to operate according to economic accountability and self-financing, to revise wage rates to encourage the reduction of time required for elaborating designs and making major decisions, and to operate at levels equal to those of their best foreign peers. Competition is to be encouraged in the development of new equipment which will raise labour productivity in forestry by 2.5–3 times by 2005. Design offices will accept only those scientific recommendations which guarantee major economic and social returns and yet still meet stern ergonomic and environmental requirements. Equipment and machinery must be competitive on international markets, particularly in the CMEA countries. Bonus payments from economic accountability funds should be used to encourage new techniques to enhance the reliability of equipment and reduce its material and energy intensity.

Goskomles proposes to use economic accountability to increase the effectiveness of testing procedures. This is to accelerate the creation of prototypes and the commencement of mass production. Low-quality and ineffective equipment will not be accepted for production. Machine-testing facilities are to develop permanent exhibitions of the latest forestry equipment and devices to demonstrate their properties and operating characteristics. Forestry-equipment builders will begin to produce comprehensive units and devices that can be modified to accommodate local conditions. Furthermore, forestry will become computerized.

Foreign scientific-technical and economic links with socialist countries are to be modified by transferring from a system primarily of trade ties to one of broad productive cooperation. Direct links will be established between enterprises and scientific-research institutes; joint enterprises will be created. Changes in the structure of foreign trade will occur through greater sales of processed wood and non-wood forest products.

Priority development of living conditions

Living conditions in forestry are at rock bottom. Fulfilment of the demands of forestry workers is 3–7 times lower than the national average according to many criteria, thus causing severe labour shortages in the sector. By 2005, forestry is to become entirely self-financing; workers are to become monetarily interested in the results of their labour and increase their productivity by 2.5–3 times; 3.9 million hectares of shelter belts are to be planted; conservation is to be extended to include 1.157 billion hectares; and the average worker's monthly salary is to rise to 370–380 roubles (apparently in values equivalent to 1989) and thereby be equal to the social norms of Gosplan USSR. In short, forestry is to measure its success in terms of its net achievements. The quality of labour is also to rise so that by 2005 not less than half the labour force has been trained in professional-technical colleges and educational centres. Some 13–14 thousand forest officers will have to be weaned away from the command-response syndrome toward more creative behaviour to facilitate accelerated economic and technical improvements in forestry. The labour problem is to be overcome by moving away from disconnected higher-educational institutions toward comprehensive educational-scientific production facilities. The entire system of professional training in forestry is to be changed by 1991 to an All-Union basis. Injuries and accidents are to be significantly reduced by as much as one half by 2005, and the number of workers doing heavy physical tasks will significantly decline for all workers, but will disappear for women.

Inspection

The existing supervision or verification system in forestry is ineffective and basically constitutes a superficial, quantitative verification of task fulfilment unsuited to future objectives. Forests' paramount environmental protective and constantly increasing social roles require forms of inspection that focus on quality and the impact of anthropogenic and other influences. Inspection should be effective and include aerial photography, ground sampling, program- or purpose-focused inspection, comprehensive and regular forest-managerial inspection and regional supervision. An extensive set of procedures is to be developed for implementing effective inspection procedures.

It remains to ensure the observance by all ministries, administrations, state, cooperative and social organizations, institutions and citizens of the established procedures of forest utilization, and rules governing forestry practices, regeneration, conservation and protection. The draft concludes with the statement that all forms of inspection will have an economic basis.

Reflections on major continuing shortcomings in the Soviet forest sector

Many problems, challenges and dilemmas in forest management extensively discussed by Soviet foresters have been summarized by Barr and Braden (1988), and placed in a concise geographical perspective by Braden (1988). Underlying myriad specific examples of forest depredation is a general inability in this sector to solve continuing shortcomings.

Distance

The forest sector appears unable to resolve Anuchin's 'geographical scissors syndrome' in which the spatial separation of resource availability and location of processing or consumption is becoming larger. Logging in the Soviet Far East, for example, is expected during the remainder of this century to shift into the middle districts of Amur oblast and Khabarovsk kray along the BAM railway to offset shortages of major commercial species in southern, settled regions. This shift will occur despite many planners' statements that commercial logging along the BAM may be cost-ineffective, even to meet export commitments, and that the BAM was an economic mistake. In short, the length of haul of timber on Soviet railways continues to grow and now is approximately 2,000 kilometres. Furthermore, in 1988 the Railway Ministry in 1988 failed to provide almost 160,000 rail cars to *Minlesprom* enterprises, despite loud protests in the country's press. The Europe-Uralia versus Siberia controversy related to timber supply for domestic markets continues to be unresolved despite the economic logic favouring regeneration and utilization of extensive deciduous stands accessible to European Soviet consumers and the elimination of expensive long hauls on the heavily used east-west trunk railways. The valid argument supporting utilization of eastern forests for strategic considerations – that is, to show the flag and demonstrate the occupancy of relatively unpopulated areas – no longer seems to have its previous currency as Gorbachev appears increasingly willing to grant Chinese, Korean, Vietnamese and Cuban loggers short-term eastern tenure, including the establishment of various joint ventures.

Shortfalls

Despite decades of apparent resolve, wood residues are inadequately used in processing as substitutes for commercial roundwood; their economic value is lost to the economy despite widespread knowledge of their intrinsic worth to national income and environmental protection. In 1988, the Soviet Union reported a shortfall in timber deliveries of 25 million cubic metres, further confirmation of bottlenecks in planning, transportation and supply. Furthermore, many *Minlesprom* enterprises are not equipped to handle wood residues and hence are forced to consume valuable timber to meet planned levels of output.

Fire

The problem of forest fires, now being documented in the Soviet press, suggests that many of the relatively accessible stands in peripheral regions have been destroyed by fire and that forest-fire suppression in important regions of Siberia and the Soviet Far East is slower to extinguish fires than in European regions (*Lesnaya Promyshlennost'*, 11 June 1988, p. 1). Surveillance is most effective in accessible forest regions. Administrative and technical shortcomings are responsible for delays in reporting and containing many of the fires (Afrin, 1990, p. 3). This situation is rendered more critical because the Soviet Union lacks a long-range forest fire forecasting ability. The Leningrad forestry research institute responsible for fire forecasting uses a technique covering one or two days (*Lesnaya Promyshlennost'*, 11 June 1988, p. 1).

Data on the number and extent of forest fires have traditionally been classified. Glasnost, however, has led these data to be declassified (Goskomstat 1989, p. 114). Although the year 1988 was only the second worst reported by Goskomstat of the four years commencing in 1985 – 24.4 thousand fires which consumed 792.4 thousand hectares of forest land and 225.4 thousand hectares of non-forest land or a total of 1017.8 thousand hectares – it had the greatest number of fires and the greatest damage to forest land. The previous two years saw extensive burning on non-forest land. In all four years, fires caused by organizations and members of the public ranged between 83 and 87 per cent of all cases. Goskomstat notes that in 1988 the burned area in Khabarovsk kray and Tyumen oblast comprised 38 per cent of the nation's total (categories of land are unspecified). The use of aircraft in forest-fire monitoring reached 69 per cent of the total area of the state-forest land in 1988 and was concentrated in the RSFSR (904.5 out of 928.4 million hectares) with aircraft surveillance.

Diffused responsibility

Nomadic logging continues apace, reflecting fragmented levels of authority for timber stands and management of growing stock. Comprehensive timber utilization cannot proceed if stand management is regularly sabotaged and perverted. Siberian stone-pine stands in the Maritime kray, for example, are at risk, not only through a history of poor harvesting practices, but also because 'logging enterprises have developed the bad habit of having several enterprises log in the same timber limits', (*Lesnaya Promyshlennost'*, 16 April 1988, p. 2). Siberian stone-pine stands comprise 6 per cent of the Soviet Union's forested area, and at least 90 per cent of its consumers could use other timber (*Ekonomicheskaya Gazeta*, No. 18, April 1988, p. 18). The species is threatened with extinction and felling should cease in the main areas where stands are found: the Gorno-Altay region, Krasnoyarsk, Khabarovsk and Maritime krays, and in Tyumen' and Tomsk oblasts. In the Gorno-Altay region, 'the fate of the pine stands remains, as before, in the loggers' hands . . . [and] the timber is logged and utilized in an excessively wasteful manner. Almost 50 per cent of the wood is lost during the logging and floating operations, and no more than 2 per cent of the entire volume of the take is used to make the main specialized assortment – pencil slats. Only one fifth of this "golden" 2 per cent is actually used in production. The rest ends up as firewood. This is how our unique national treasure is being wasted. The worst thing of all is that this type of timber utilization has become an accepted practice.' Changes in operating procedures clearly could enhance logging productivity and reduce forest depredation.

Furthermore, air pollution from heavy industry even at Bratsk since 1986 has killed over 80,000 hectares of pine forests surrounding the city's communities: 'the damage to the environment around Bratsk and other industrial centres in the Angara river area has been so severe that it is impossible not to see its consequences' (*Lesnaya Promyshlennost'*, 2 August 1988, p. 1). Pollution in such cases further exacerbates the damage caused by predatory harvesting practices. Bratsk city council has asked that construction of the viscous pulp unit at the Bratsk LPK be stopped (this capacity would replace some of that to be closed at the Priozersk mill on Lake Ladoga and at the Baykalsk mill – *Sotsialisticheskaya Industriya*, 21 February 1988, pp. 1–2). The newspaper reports that an average of 170,000 metric tons of pollutants are released into the air annually (presumably by the LPK and aluminium plant) at Bratsk: concentrations exceed the maximum permissible levels by 50 times for methylmercapton, 13 times for carbon disulphide, 4.5 times for hydrogen sulphide and 3.5 times for hydrogen fluoride.

Underfunding

Placing responsibility for harvesting and regeneration in one ministry on the surface is a good idea if the requisite funds are provided by management for silviculture. The press suggests, however, that not only are forestry activities still regularly underfunded, but delays have occurred in the transfer of enterprises from *Goskomles* to *Minlesprom*. Debate also continues on the possibility of enterprises being allowed to declare bankruptcy, something that still appears at odds with the goals of central planning and a system that does not encourage profitable enterprises to fill the timber orders of ineffective enterprises autonomously and effectively. Gorbachev has not yet instituted price reform, which will be necessary before either central planners or enterprise managers can make better investment choices in the allocation of resources. Furthermore, he has not addressed the issue of regional differentiation in allowable costs and revenues; neither has he developed a mechanism for continuous adjustments in prices to reflect market choices and vicissitudes in supply and demand. Even if prices are revised, they will quickly lose their significance if they do not serve as current and effective signals to decision-makers. We may speculate that numerous logging enterprises that are already operating at a loss will be subject to even greater outlays if they receive inadequate compensation for forestry and regenerative activities.

Regional forest degradation

When the Soviet Union's 20 major economic regions are ranked from worst to least according to extent of damage to forests (Goskomstat, 1989, p. 8) degradation appears to be particularly severe in those regions with the country's major forests: Eastern Siberia (1), North (2), Urals (3), Western Siberia (5), Northwest (7), Far East (8) and Baltics (9). Two regions lacking major forests also register severe damage from pollution: Donetsk-Dnieper (4) and the Centre (6). Forests in only three current or potentially important forest regions are ranked relatively low in pollution damage: Volga Littoral (12), Belorussia (13) and Volgo-Vyatka (18). The significance of these statistics is that they show harmful effects to the forest in the country's major forests, in regions of major national timber supply, and in European-Uralian regions of great potential significance to the Soviet Union if scientific forestry practices, including effective sustained yield-management and silviculture and greater use of deciduous species, gain wider acceptance.

Pollution

Significant amounts of timber are still floated on the country's rivers and lakes, including Lake Baykal, despite enormous annual and accumulated losses (*Lesnaya Promyshlennost'*, 28 July 1988, p. 2; 25 August 1988, p. 2). Drift-floating is still a major activity in the Soviet Union, although it is to continue to be progressively reduced through the 1990s. Towing rafts of wood across Lake Baykal is to cease by 1995. In the latter case, ships already exist to transport the wood but loggers balk at the cost – 40 kopecks per cubic metre by raft, 2 roubles per cubic metre by barge (*Lesnaya Promyshlennost'*, 25 August 1988, p. 2). A minimum of 757 thousand cubic metres were to be moved by water across the lake in 1988 (levels in recent years have exceeded 1.2 million cubic metres). Furthermore, the pulp and paper operations at the Baykalsk and Selenga mills continue to pollute Lake Baykal, and the date for the closure of the former and cessation of waste-water discharge of the latter now seems to be 1993 rather than 1991. Likewise, compensatory capacity at Ust'-Ilimsk and Bratsk for the conversion of the Baykalsk mill into a furniture factory and the closure of some capacity at the Priozersk mill on Lake Ladoga is unlikely to be installed on time – and in fact may be inadequate to stop pollution of the lower Angara River. Continuing complaints in the Soviet press lead to the regrettable conclusion that, if pollution of Lake Baykal and its environs does not soon stop totally, that body of water will become another major ecological disaster like the Aral Sea; although the lake will contain water, its quality will have been irreversibly damaged.

Goskompriroda (1989, p. 105) reports that principal harvesting has been prohibited in a protective zone around Lake Baykal and that the allowable cut in the river basins around the lake has been reduced in Buryatia by 18 per cent, in Chita oblast by 25 per cent and in Irkutsk oblast by 80 per cent. NKSSSR (1987, pp. 574–5), however, reports that polluted water from the Baykalsk pulp and paper plant dumped into Lake Baykal constituted 34 million cubic metres (32 per cent of the polluted flow in the region), and the flow from the Selenga pulp and paperboard mill constituted 13 million cubic metres or 12 per cent of the total polluted flow. These two mills thus comprise nearly half the polluted effluent in the region. Correction of the problem is far behind schedule for the Baykalsk mill and for other enterprises in the Baykal basin. NKSSSR (1988, p. 248) reports that the Sloka pulp and paper mill at Yurmala in Latvia annually dumped 15 million cubic metres of polluted effluent into the Baltic in 1985, 1987 and 1988. An extensive article on the implications of closing the Sloka mill suggests that it has apparently been shut down (CDSP, 42 (4), 1990, pp. 10–12) but for how long and at what direct and indirect cost is not clear. The Baykalsk mill was to be closed in 1990 and the Selenga mill converted to a closed-water system. It is unclear, however, whether *Minlesprom* is willing or able to carry through these promises or to commission replacement capacity for closed mills, including the Priozersk pulp and paper mill on Lake Ladoga (Lavrov 1989, p. 674), at Bratsk and Ust'-Ilimsk, either on time or in face of local environmental opposition in both replacement centres.

'The Baykal syndrome'

Lake Baykal continues to deteriorate (Massey Stewart, 1990). Pollution stretches for over 20 square kilometres around the Baykalsk mill, despite the plant's purification facilities (*Lesnaya Promyshlennost'*, 28 January 1988, p. 2). The numbers of endemic algae in the lake have dropped by half and include the Baykal Melozira, the main link in the omul's (a member of the salmon family) food chain. Green algae has spread since 1970 throughout much of the lake. Omul roe is being massively destroyed along the Selenga river. The damage originates with the two pulp and paper mills whose activities are not to be significantly altered before 1993. Compensatory facilities, a dissolving chemical pulp mill at Ust'-Ilimsk and a viscous pulp mill at Bratsk, are being built (*Lesnaya Promyshlennost'*, 26 July 1988, pp. 1–2) to replace Baykalsk output (this mill is to adopt alternative production which is 'ecologically pure – it is to become a furniture and furniture-assembly plant' – *Sotsialisticheskaya Industriya*, 21 February 1988, pp. 1–2), and the Selenga pulp and paperboard plant is to adopt a closed-water cycle (already developed years ago) for which 'there is no 100 per cent guarantee that it will work'.

Academician Boris Laskorin, chairman of the Commission of Water Conservation Problems, notes (*Lesnaya Promyshlennost'*, 26 July 1988, pp. 1–2) that when the Baykalsk mill was authorized, no integrated approach to the environment was followed. Furthermore, the mill was already operating when it was discovered that its wood base was unsuitable for super-chemical pulp. The mill has been receiving timber from Khabarovsk and Krasnoyarsk krays for the past twenty years. The Baykalsk operations are being transferred to Bratsk and Ust'-Ilimsk but the timber will continue to be supplied from these other regions. Lake Baykal water also proved to be unsuitable for rinsing and a special purification process was undertaken which will not be transferred to the two Angara mills: these in effect will use Baykal water, patently unsuitable for this type of pulping. The Angara mills are also not diluting their effluent 1,000 fold as required because it is neither practical nor cost-effective. Hence the pollution dilemma at the Baykalsk mill has simply and finally been transferred downstream. Until effective changes are made, therefore, the mills will continue to harm the lake and the Angara river. Given normal delays in commissioning new capacity in the Soviet Union, their operation may well continue through the 1990s. Furthermore, accidents are frequent at the Baykalsk mill and hence even if construction of the pipeline to discharge effluent into the Irkut river should proceed, the Angara itself will still be harmed by all proposed changes.

An interview with the then Minister of *Minlesprom*, M. I. Busygin, published in *Sotsialisticheskaya Industriya* (21 February 1988, pp. 1–2) does not offer convincing reassurance that damage to the lake will soon cease. Busygin, while trying to sound accommodating, became upset when challenged on continuing pollution to Lake Baykal while new facilities are built elsewhere, stating that the public's 'ecological hysteria' is obstructing the work of engineers and directors, and perhaps unwittingly adding that 'protecting the environment is definitely not the highest priority of directors at a number of enterprises. We are no longer cajoling or admonishing these people, but are expecting them to take on the full measure of their responsibility.' He rejects calls for the Baykalsk mill's immediate closure as 'an untenable solution which could damage the economy enormously' but admitted that public outcry did secure closure of the Priozersk mill because it 'had no purification facilities at all. The Baykalsk mill is equipped with a unique technical system unparalleled anywhere else in the world.'

Underlying some of Busygin's attitude is his realization that the Soviet Union's demand for paper is being met on average by approximately one half (it ranks forty-two in the world per capita), and that his industry is also causing severe pollution and ecological damage. Busygin notes that capital investment in this sector increased by only 28 per cent in the 1970s compared to 210 per cent in the 1960s and that 'many enterprises are equipped with horrifically out-of-date machinery and only 5 per cent of the units and machines now in operation are up to world standard; . . . we have to retool enterprises which are 80 to 100 years old and to build new ones'. Busygin states that the country would have to spend one billion roubles to increase per capita consumption of paper by just one kilogram, that is from the present average of 36 kilograms to 37 kilograms. The industry is seeking to raise these levels 'in the near future' to reach between 60 and 70 kilograms and produce 15–16 million tons of paper and 11 million tons of paperboard annually (the current combined total is 10 million tons and the country has dropped to fifth place in total output after China). Rebuilding and retooling will permit an increase of only 5 to 15 per cent in capacity, hence growth will come at new plants.

Only one third of the Soviet Union's pulp and paper enterprises have purification facilities, and Busygin does not foresee that their cost can be justified at many enterprises. Hence some operations were to close: three old mills in Archangel oblast by 1991; the Priozersk mill in Leningrad oblast is already producing other products; the Okulovka pulp and paper combine in Novgorod oblast, and the Lyaskelya pulp and paper combine in Karelia no longer produce chemical pulp, alcohol or yeast. Busygin states that clean-up of Lake Drozdovo will soon commence where the Priozersk pulp mill previously dumped its unpurified effluent.

Inferior status

Lesnaya Promyshlennost' on 10 March, 1988 (p. 2) published a lengthy article describing the depredation of Tyumen forests that has accompanied the well-known development of the region's oil and gas industries. The key point of this 'lament' is that the foresters' expectations for the region's development have not been met: 'With hindsight they can see that they were right off track from the very beginning.' The area of land allocated permanently or temporarily to the energy sector equals that logged each year for the principal cut. The forests everywhere, however, are being destroyed 'for no apparent reason'. Chemicals, effluent, industrial and construction debris of all sorts, and abandoned logs litter the landscape: 'an unregulated process of degradation of the northern taiga is underway . . . the oil and gas industry has no ecology experts, no anti-pollution devices or control measures . . . Leakage of oils, drilling solutions and mineralized water is numbered in the hundreds of thousands, if not millions, of cubic meters and tons . . . The scale of this disaster increased in proportion to the expansion of drilling operations, the number of fields and the length of the pipelines.'

The article cites extensive leaks, spills and environmental degradation, including damage to local inhabitants. Power-line rights of way have been cleared with bulldozers and large spoil-heaps left which are both fire hazards and breeding grounds for forest predators. Although the energy sector has its own environmental protection agencies, they do not exert an influence on exploration and production practices, many of which use environmentally inappropriate technology.

Forestry is understaffed and has a high labour turnover. Only 67 foresters work in the 18 million hectares of state forest in the four forestry operations on the oil-rich right bank of the Ob'. The forestry staff handling land management, registering surveys and supervising energy operations is inadequate and is poorly paid. Many planning, inventory and cartographic materials are obsolete; furthermore, only one third of the state forest has been charted for forestry management in Tyumen oblast. Forestry enterprises in the middle Ob', which has one fifth of the state forest area and most oil extraction, are effectively equipped for the pre-oil era. Exploration crews are accused of damaging the environment indiscriminately and simply paying fines with state funds, even though the fines are much larger than the forestry budget. In general terms, then, this region has incurred an ecological disaster in its forests and waters.

Primary industry is wreaking havoc in other regions as well. Laskorin observes (*Lesnaya Promyshlennost'*, 26 July 1988, pp. 1–2) that because of the processing of sulphitic ores in metallurgy (Noril'sk) and the use of sulphurous coal (Cheremkhovo and other power stations, including those of *Minlesprom*): 'large expanses of the taiga are dying as a result of the millions of tons of sulphurous anhydride which are emitted annually into the atmosphere around Noril'sk . . . Soviet design work on gas scrubbers is not going at all well, . . . the economic planners do not want to make major capital investments in the purification field. They think only of the immediate costs, without considering the benefit to the national economy as a whole.' Laskorin cites the extensive world experience with gas purification facilities but notes that the Soviet Union is only at the decision-making stage at present. 'We are sadly lagging in the implementation process.' Laskorin states that attempts first to deal with ecology and then with industrial development are hardly successful: 'We are only now beginning to deal with this, but as usual it's more a question of words than deeds.' Despite an agreement to study water conservation, make ecological recommendations and correct past mistakes, reached in 1987 between the Academy of Sciences and *Minlesbumprom*, the latter: 'pushed this program to one side. Our country still profits from the knowledge that we are very rich in forest resources. That is why our forest management and timber utilization practices have been so wasteful. Our traditional approach is based on the idea (not unique to the Soviet Union but common to all mankind) that natural resources are inexhaustible.' When asked if such an attitude did not begin to break down about twenty years ago, Laskorin replied that 'it is still very much in evidence, although perhaps it exists more in the subconscious.' Citing examples of the eutrification of lakes based on faulty policies, Laskorin observes that 'if the discharge of pollutants is not stopped, our rivers and lakes will contain nothing but effluents' especially in the Priozersk region of Lake Ladoga and the critical situation of the Baltic coast and Amur river.

Attrition

Lesnaya Promyshlennost' on 9 August, 1988 (p. 2) published data on wastage of marketable timber by *Vologdalesprom*, a territorial production amalgamation. Referring to this waste as 'attrition', the article demonstrated that 20.8 per cent (2,475,700 cubic metres) of the marketable volume to be harvested in 1986 was lost through these activities:

Undercut (shortfall in harvest)	527,000
Discarded at felling site	727,900
Left at loading areas	352,100
Lost in transport to lower wood yard	140,000
Lost at lower wood yard	728,700
Total lost	2,475,700 cu. m.

Similar losses in 1986 for the amalgamation *Tomlesprom* comprised 12.1 per cent (778,600 cubic metres) of the marketable volume to be harvested, but Bratsk LPK in 1987 lost only 1,700 cubic metres in these categories. Until 1988, it appears that such losses were tolerated but that now logging enterprises are responsible for all timber volumes commencing with felling activity. Nevertheless, until now considerable expense has been incurred by Soviet loggers in harvesting timber that never reaches consumers. Such activity further exacerbates shortages of labour and equipment, and probably accounts for the much touted deficit of 25 million cubic metres in roundwood reported in 1988 (*Lesnaya Promyshlennost'*, 7 February 1989, p. 2). *Minlesprom* itself failed to obtain six million cubic metres (*Lesnaya Promyshlennost'*, 26 January 1989, p. 1), partly due to the failure of the Ministry of Railways to provide 126,000 rail cars in the first eleven months of 1988 (156,800 by the year's end) (*Lesnaya Promyshlennost'*, 15 December 1988, p. 2; 19 January 1989, p. 2).

It seems ironic that the Soviet Union should consistently experience shortages of timber while foresters write that through better organization of activity, the country (in 1989) could obtain an additional 10 million cubic metres of roundwood and foreign customers at least an additional 7 million cubic metres. *Lesnaya Promyshlennost'* (7 February 1989, p. 2) has pondered what actions should be taken to overcome timber shortages:

increase harvesting? take a crash-course approach in building new enterprises? intensively wring out the forests of Siberia and the Far East? The recommended approach, however, is to introduce on a wide scale resource-conserving technology that will more fully and efficaciously utilize secondary wood materials. Even simple calculations suggest that currently some 12 million cubic metres of wood residues remain unused at forest sector enterprises and an even larger amount of timber is used for fuel. Another analyst (*Lesnaya Promyshlennost'*, 15 December 1988, p. 2) reminds readers that 95 million cubic metres, one quarter of the annual timber harvest, is consumed as fuel in the Soviet Union. Nevertheless, if wood residues from logging and processing were utilized, and if sheet products and paperboard were substituted for much processed wood, then 'it would not be necessary to pull down our forests indiscriminately but to conserve them for our descendants and significantly improve the general ecological situation' (*Lesnaya Promyshlennost'*, 7 February 1989, p. 2). The same issue of the newspaper reports that the Soviet Union annually produces 77 million cubic metres of wood residues at its enterprises, of which 35 million cubic metres are processed and 9.5 million are dumped and burned. In addition, 20 million cubic metres of roundwood are lost through rot. The reasons are not unusual: 'current pricing policies are ineffective to stimulate additional use or conservation of these materials, and the supply of equipment and chemicals is inadequate . . . By the year 2005, however, 23 per cent more timber will be extracted from the taiga than now, but still will not meet the country's need for timber.'

Cumulative induced diseconomies

In light of the current situation in the Soviet commercial forest, the title of an article by Dr V. Sukhikh, vice-chairman of *Goskomles*, in *Lesnaya Promyshlennost'* (17 December 1988, p. 2) – 'Do we have enough forests?' – sounds close to the title of *The Disappearing Russian Forest: A Dilemma in Soviet Resource Management* (Barr and Braden, 1988). Sukhikh states that in principle forestry in the Soviet union could be intensively conducted on 100-150 million hectares in western, southern and central regions to provide annually 400–500 million cubic metres. But the basic truth lies in the fact that the country has abundant forests and harvesting occurs anywhere and in any amount. Forestry is practised on 792 million hectares, of which 385.1 comprise commercial timberland with 47.1 billion cubic metres of growing stock (74 per cent of the volume is coniferous). The allowable annual cut is 636.2 million cubic metres but this figure rises to over 700 million if improvement felling, weeding and brushing is included. In 1987 the country had a principal cut of 339 million cubic metres. Improvement felling produced 68 million cubic metres and 15 million cubic metres was felled on collective and state farms and by other administrations for a total cut of 422 million cubic metres. Thus, according to Sukhikh, existing raw materials if correctly exploited can satisfy the economy's need for timber. His evidence, however, could equally lead to his main question being answered, 'no': because the country does not efficaciously exploit its forests, at least half the timber seems to be supplied at an economic loss even in Soviet accounting terms.

Thus Sukhikh is really saying that Soviet forestry and forests are a mess. Sukhikh's assessment is similar to that of Barr and Braden. He states that the most productive and high-quality commercial forests of Europe-Uralia, adjacent to the Siberian railways and in the Far East have been depleted and cannot meet the country's growing need for timber. Remote forests require excessive amounts of capital for development. The essence of Sukhikh's argument is that the accessible forest has been destroyed through a combination of inaccurate data bases and *goszakazy* which are at odds with timber limits' capacities. Losses of allotted timber at the cutting area comprise on average up to 10–15 per cent in Europe-Uralia, 15–20 per cent in Siberia and 20–25 per cent in the Far East. Furthermore, Sukhikh observes that adequate Shade Intolerant Hardwood Species (SIHS) timber and Group I stands ('protection/ forests/sanitary and selective cutting only') exist in Europe-Uralia to provide a clear 'yes' to his basic question, while still recognizing that timber supplies are not inexhaustible. The solution is to use them effectively. While many speak of taking a comprehensive approach to forest management, the change of 1988 in which *Minlesprom* became responsible for regeneration was meant to ensure comprehensive utilization. In fact, states Sukhikh, the situation deteriorated in 1988: overcutting in many regions has increased and much falsification has crept into official reporting on forestry activities. The solution he feels lies not in comprehensive use alone, but in a more thorough scientific approach to all aspects of timber management, including better assessment of the forest and exclusion of many types of area from the designation of forested land. Approximately 40 per cent of the state-forest reserve was measured 35–40 years ago with crude airphoto techniques. In some places, where the forests have been reassessed, their volume in fact is overstated by 1.5–2 times. To Sukhikh, the solution to effective (i.e. sustained-yield) management is an overall professional approach to tasks and adherence to well-prepared plans supported by realistic allotments, forest valuation, and monetary evaluation, and supervision of activities through aerial and satellite verification. Equally important is the provision of adequate levels of equipment and technical support in all branches of forest utilization, including that related to SIHS, to low-quality wood, and to residues. Specialists in 'forestry and the forest-product sector will have to work together and not permit departmentalism to prevent all the resources of the biosphere from enhancing the eternal well-being of humankind.'

Conclusions: priorities, constraints and contradictions

Many of the problems of the Soviet forest sector have survived numerous earlier revisions, including the 1988 restructuring of the forestry and forest-product sectors of the Soviet Union (Barr, 1989), and remain today. Others have become exacerbated; none seem to have been resolved. Gorbachev's restructuring is meant to strengthen the economy, but events suggest that with the weakening of central administration have come problems of coordination that threaten the coherence of the industrial system. The need for reform has probably never been more acute, but restructuring appears to have caused managerial confusion and perhaps even added to environmental problems without resolving those issues that initially prompted the attempts at change. Numerous sources stress that problems should be solved through more effective central administration and thereby confirm my conclusion that restructuring may succeed in the long-run only if Soviet and Russian traditions are incorporated into new approaches using modern technology for administration and supervision. Gorbachev's and the first post-Gorbachev era's perestroyka seems based on the expectation that managers and others will suddenly start to behave as if they have a long tradition of decentralized administration and decision-making more in keeping with Western experience. Current approaches seem strangely out of keeping with many of the positive aspects of Soviet experience and behaviour and hence are unlikely to succeed. In the meantime, the Soviet forest and environment continue to suffer severe depredation and degradation.

Geography, attempts at restructuring, conceptualizations of the future, on-going systemic shortcomings and regional industrial composition of the forest sector should be compatible if forces ensuring equilibrium were at work in the Soviet political economy. Reform to date, however, has not ensured that equilibrating processes should be permitted to form and then to interact over space in the allocation of resources, the determination of timber utilization and the formulation of environmental policy. Although broad patterns of regional localization in timber-harvesting and types of wood-processing can be ascertained, numerous characteristics of forestry and the forest-product sector suggest that myriad inefficiencies or dysfunctional organization and behaviour impede the realizations of effective utilization of geographical space, growing stock and the environment. Although the broad goals of sustained yield-management are understood in the Soviet Union, logging practices, regional harvesting priorities and resource mismanagement continue to induce cumulative diseconomies in the quality of growing stock and future options for regional resource utilization. The Soviet literature leaves little doubt that misguided priorities and wasted opportunities have created a situation today in which much of the so-called accessible and commercial forest has undergone and continues to undergo depredation, leading to the prospect that stands in many regions are likely to lose their economic significance in the near future.

Thus the fate of many regional forests and the viability of settlement and economic systems dependent on them is bleak. Some of the cause for this dilemma lies in how the growing stock has been misused and wasted by the forest sector and myriad non-forestry ministries, including the lack of effective regeneration and destruction of the reproductive environment. Thus the immediate environment in which trees stand and grow has been damaged by unsuitable harvesting practices. Operations within the forest sector continue to harm not only the forest but also non-forest environments like rivers and lakes, fish and animal habitats and ecologically sensitive terrains in harsh environments that are slow to heal or regenerate.

The forest environment, however, is also being adversely affected by atmospheric and environmental pollution from other sectors of the economy and by competing land-uses in areas adjacent to urban and industrial activities. Large expanses of growing stock, however, continue to fall victim to horrendous levels of atmospheric emission by smelters and smoke-stacks of other activities, and to extensive fires, especially in more sparsely populated areas where detection and suppression facilities are poorly developed.

Effective restructuring should lead to enhanced recovery of value per unit of input and should reduce the overall levels of timber harvested, especially in peripheral and marginal forests. Greater efficiency in resource utilization will alleviate pressure on European forests currently undergoing depredation from overcutting by enhancing the value of presently unpopular species, and by ensuring greater recovery per unit area. Furthermore, if the real costs of transportation were allowed to influence locational and species' choices in logging and wood-processing, both the scale and structure of local or regional production and consumption in the Soviet Union should display greater efficiency, particularly if regionally differentiated costs of the factors of production were incorporated into the production strategies of each enterprise.

The Soviet economic system still appears to place heavy emphasis on achieving production goals at any cost, although costs based on effective prices cannot be calculated without the much-delayed implementation of price reform. Perestroyka to date has not produced an effective price mechanism with which harvesting and regeneration alternatives can be assessed, species utility evaluated, regional utilization priorities assigned and spatial choice exercised in the allocation of industrial investment capital. Regional development options available to *Minlesprom*, to territorial and local organizations, and to joint Soviet-foreign enterprises are increasingly being pre-empted by a lack of comprehensive reform throughout the national political, economic and legal system, the rapid rise since 1988 of the 'green movement' among myriad legislative and independent bodies, the depredation of much of the accessible European, Uralian and central Siberian forest and the widespread destruction of the forest from atmospheric and other human-induced pollution, including radioactive emissions from Chernobyl.

In sum, without true restructuring of the political and economic system of the Soviet Union, and provision of all the attendant requisite changes in operating procedures, neither the forest nor the environment will stand a chance of being protected and their integrity safeguarded. Not only must the economic system change, but social and cultural values and institutions truly intolerant of loutish environmental mismanagement must be established and internalized by the population and allowed to regulate and control those responsible for the forest and the environment. True restructuring should permit effective choices to be made in the regional use of raw materials and the recovery of as much value per unit of input as possible. True restructuring should lead to protection of the environment if costs of production are allowed to determine levels and locations of output and if environmental legislation is enacted and enforced. No set of measures, however, will remove present patterns of regional dysfunction and environmental degradation unless Soviet society at all levels displays the will to ensure that the present, largely self-induced constraints on systemic improvement are overcome and that the rampant contradictions among Soviet priorities, institutional goals and industrial practices are eliminated.

Suggested reading

Barr, B.M. and Braden, K.E. (1988), *The Disappearing Russian Forest,* Rowman & Littlefield, Totowa, N.J.

Dienes, L. (1987), *Soviet Asia: Economic Development and National Policy Choices,* Westview Press, Boulder and London.

Pryde, P.R. (1991), *Environmental Management in the Soviet Union,* Cambridge University Press, Cambridge.

Weiner, D.R. (1988), *Models of Nature: Ecology, Conservation, and Cultural Revolution in Soviet Russia,* Indiana University Press, Bloomington and Indianapolis.

Ziegler, C.E. (1987), *Environmental Policy in the USSR,* University of Massachusetts Press, Amherst.

References

Afrin, M.I. (1990), 'Lesnye pozhary Sakhalina: stikhiynoye bedstviye?' *Lesnaya Promyshlennost'* (monthly journal) 6, p.3.

Barr, B.M. (1989), *'The Soviet forest sector now and in the future: too much glasnost, too little perestroyka?'* The H.R. MacMillan Lectureship in Forestry, Vancouver, B.C.: The University of British Columbia, Faculty of Forestry, 19 January.

Barr, B.M. (1990), 'Forest and fishing industries', in Rodgers, A. (ed.) *The Soviet Far East: Geographical Perspectives on Development,* Routledge, London.

Barr, B.M. and Braden, K.E. (1988), *The Disappearing Russian Forest,* Rowman & Littlefield, Totowa, N.J.

Braden, K.E. (1988), 'Environmental issues in Soviet forest management', *Soviet geography,* 29, 6, pp. 599–607.

CDSP (*The Current Digest of the Soviet Press*), (weekly publication), various issues, Columbus, Ohio.

Ekonomicheskaya Gazeta (since 1 January 1990 published as *Ekonomika i Zhizn'*), (weekly publication), various issues, Moscow.

Goskompriroda (1989), (Gosudarstvennyy Komitet SSSR po Okhrane Prirody 1989), Doklad. *Sostoyaniye priorodnoy sredy v SSSR v 1988 godu,* Moscow.

Goskomstat (1989), *Okhrana okruzhayushchey sredy i ratsional'noye ispol'zovaniye prirodnykh resursov v SSSR. statisticheskiy sbornik,* Goskomstat, Moscow.

Lavrov, S.B. (1989), 'Geoecology: theory and some practical issues', *Soviet geography,* 30, pp. 670–9.

Lesnaya Promyshlennost' (re-named *Lesnaya Gazeta* 14 August 1990), (newspaper, published thrice weekly), various issues, Moscow.

Lesnoye Khozyaystvo (monthly publication), VO 'Agropromizdat', Moscow.

Massey Stewart, J. (1990), 'The great lake is in peril', *New Scientist,* 126, 1723, 30 June pp. 58–62.

Medvedkov, O.L. (1990), *Soviet Urbanization,* Routledge, London.

Micklin, P.P. and Bond, A.R. (1988), 'Reflections on environmentalism and the river diversion projects', *Soviet Economy,* 4, 3, pp. 253–74.

NKSSSR 1987 (*Narodnoye Khozyaystvo SSSR v 1987 g.*), (1988), Finansy i Statistika, Moscow.

NKSSSR 1988 (*Narodnoye Khozyaystvo SSSR v 1988 g.*), (1989), Finansy i Statistika, Moscow.

Panel on the state of the environment at the start of the nineties (1990), *Soviet geography,* 31, 6, pp. 401–68.

Pryde, P.R. (1991), *Environmental Management in the Soviet Union,* Cambridge University Press, Cambridge.

Sagers, M.J. (1988), 'News notes: New state committee for environmental protection created', *Soviet Geography,* 29, 6, pp. 635–6.

Sagers, M.J. (1989), 'News notes', *Soviet Geography,* 30, 8, pp. 684–5.

Sotsialisticheskaya Industriya (daily newspaper), various issues, Moscow.

'Soviet joint ventures: developments through to the first quarter of 1990', (1990), *PlanEcon Report,* 6, 17, p. 14.

United Nations, Economic and Social Council, Economic Commission for Europe, Timber Committee, (1989a), *Timber trends and prospects in the USSR,* GE.89–45791/1003B, 19 September, Geneva, pp. 25.

United Nations, Economic and Social Council, Economic Commission for Europe, Timber Committee, (1989b) *Outlook for forests and forest-products sector of USSR,* ECE/TIM/48, Geneva.

Ziegler, C.E. (1987), *Environmental Policy in the USSR,* University of Massachusetts Press, Amherst.

8

Perestroyka and the Soviet transportation system

Robert N. North

ВЛ10-1410

Introduction

In the space of a few months in 1986 and 1987, perestroyka became one of the two best-known Russian words in the non-Soviet world. Was it merely the latest in a series of buzzwords, temporarily fashionable in the rhetoric of economic reform but signifying little real change and brought to international attention largely by the charisma of Mikhail Gorbachev? Or did its widespread currency reflect a genuine revolution in the Soviet economy? The public disillusionment which set in as the decade ended suggested the former. However, evidence from the transport sector suggests that perestroyka should not be lightly dismissed. At the early stages of reform, transport organizations were forced to confront radical changes in their operating environment. Their reactions had both positive and negative effects on the transport process and it became clear that if perestroyka continued, it had the potential to bring even more radical changes – in the market for transport services, in investment policies for transport, in the interrelationships among transport modes and in their relationships with other sectoral and regional organizations. These changes in turn would affect the geography of Soviet transport facilities and traffic flows. It also became clear that if perestroyka failed to rejuvenate transport, the consequences for rejuvenation of the economy as a whole could be catastrophic.

If the initial stages of perestroyka demonstrated great potential, its evolution in 1989 and 1990 gave cause for apprehension. By mid-1990 it appeared to have reached or to be reaching the limit of its impact in transport, barring some sharp new impetus. This was partly because the rewards for extra effort at the lower levels of the transport hierarchy had begun to stagnate or decline, and partly because the traditional bureaucratic, centralized administrative system was successfully reasserting itself and nullifying the more radical aspects of change. The further measures for economic reform, approved by the USSR Supreme Soviet in October 1990 after months of acrimonious debate, seemed unlikely to change this situation. Steering an uneasy course between the gradual and precipitate approaches to marketization advocated by rival economists, they appeared to leave ample manoeuvring room for the bureaucrats. Indeed it was becoming apparent that the most powerful forces for economic change in general, and change in transport in particular, would probably be nationalism and the drive for regional economic autonomy – products of glasnost rather than perestroyka. At the same time there loomed ever-more menacingly the spectre of power shifting to conservative forces inimical to the whole reform process. In transport that would mean a reversion to old problems and inefficiencies.

This chapter will firstly outline the salient features of Soviet transport in the pre-Gorbachev era. Secondly, it will describe certain problems, of particular interest to geographers, that faced the system in the last years of that era and the solutions attempted or proposed. Thirdly, it will assess the impact of perestroyka on transport so far, with particular reference to the geography of transport facilities and traffic flows. And fourthly, it will try to assess the long-term prospects for transport under a variety of scenarios for the future of Soviet society and the Soviet economy.

Soviet transport before Gorbachev: systemic features

Investment policies and practices

Though precisely comparable figures do not exist, it appears that transport's share of investment in the Soviet Union differs little from its share in many developed Western countries. The similarity, however, may be coincidental, since the various influences on transport investment do not carry the same relative weight in the two cases. The outcome of investment appears to differ in that much of Soviet transport always seems to be operating close to the limits of capacity.

Several factors tend to depress transport's share of investment in the Soviet Union by comparison with its share in capitalist countries. Firstly, in capitalist economies transport is often used as a development leader: investment in transport runs ahead of demonstrated demand, in the hope that the existence of facilities will stimulate economic development in general and the demand for transport in particular. By contrast in the Soviet Union, transport facilities are normally built only to meet demonstrated demand with the least possible investment. This is justified on the grounds that if all development is planned there should be no need to stimulate it. Not only is surplus capacity therefore less likely to be found than in the West, but transport-using activities are less likely to be stimulated and some latent demands for transport may not be identified at all.

Secondly, some transport investments in capitalist countries fail to have the expected economic impact, so the capacity they create remains permanently under-used. This was the case with the more northerly transcontinental lines in Canada, which were eventually nationalized because the traffic generated could not pay for the investment. Also, some investments are out-competed before they are depreciated. This was true of many canals in both Europe and North America when railways were built. On the other hand successful investments often create demand quickly, necessitating further investments to ease congestion. This has been the case with much road-building. Consequently, capitalist countries tend to have both surplus transport capacity and extreme pressure on popular modes and routes. In the Soviet Union, by contrast, there has been pressure on consumers to distribute their custom among available facilities to minimize the underuse of investment. There may be higher costs elsewhere as a result.

Thirdly, traditional Soviet economic thinking has held that transport produces no material goods, so it should have lower priority for investment than sectors of the economy that do. In an economy where government priorities far outweigh market forces, such an attitude is bound to depress investment. In more recent times the argument has been offset to some extent by the recognition that transport creates 'place utility'.

Fourthly, the idea that surplus capacity and competition are desirable to promote efficient service found little favour in the pre-perestroyka Soviet Union. On the contrary, all transport modes were continually expected to seek new ways of meeting demand through reorganization and heightened discipline rather than investment. They were also expected to cooperate rather than compete: the model of Soviet transport purveyed both to its trainees, through countless textbooks, and to the outside world was one of efficient and harmonious integration – the 'unified transport system of the Soviet Union'.

Soviet efforts to build an efficient transport system with minimal investment have not always had the intended effect. Firstly, planners have persistently failed to recognize that transport construction can stimulate traffic even in a comprehensively planned economy. The lesson seems hard to learn. East of the Urals, for example, both the South Siberian railway in the 1950s and the Tyumen-Surgut-northwest Siberian railway in the 1970s and 1980s were soon overloaded by unanticipated traffic. Since adding capacity shortly after construction can be more expensive than incorporating surplus capacity in the first place, a policy of minimizing transport investment can sometimes have the opposite effect in the longer term.

Secondly, the 'unified transport system of the Soviet Union' has been a concept honoured more in the breach than in the observance. Transport has always been organized on strictly modal lines, with a federal or republic ministry controlling each mode. In an economy permanently plagued by shortages and with each economic unit judged on its ability to meet and surpass output targets, the modal ministries and their subunits have been forced to compete rather than cooperate. They have competed for investment, supplies and traffic. Even where cooperation could have produced gains for both partners it has rarely occurred, since the economic and organizational environments have promoted mistrust and a preference for self-sufficiency. In consequence shipments by more than one mode have often received poor treatment and customers have learned to avoid them. Container and other forms of multimodal traffic have been slow to develop. This suggests that investments have been used inefficiently, since better intermodal technology has been a major source of cost reduction in the West.

Thirdly, an administered economy creates its own imbalances, which in transport can waste investments and create surplus capacity. A common shortcoming of Soviet water transport, for example, has been underinvestment in shore facilities relative to that in ships. The two types of investment have been controlled by different government organizations and levels in the administrative hierarchy which do not coordinate well. Ships are underused because they spend much of their time waiting to load and unload.

The interpretation of Soviet principles for transport investment has changed over the years. Until the 1950s the desire to minimize investment led planners to concentrate on one mode. The railways were chosen as the best all-purpose mode for Soviet conditions; for many years they almost had a monopoly of domestic transport over all but the shortest distances. In the early stages of Soviet industrialization this was probably sensible, since it facilitated a concentration of effort and scale economies in both transportation research and the production of equipment. But by the 1950s it was evident that the country would be better served by matching traffic to the modes best suited to carry it (see Figure 8.1). Since then, investment has been spread more evenly (Table 8.1). Pipelines have taken over crude-oil and natural-gas transport; much long-distance passenger traffic now goes by air; and short-distance freight has shifted to motor transport, albeit more slowly than in most countries. These changes in policy have necessitated heavy remedial investment in previously neglected modes. Rural and trunk roads are the most striking current example.

Another change in the past forty years is that as the country has played a greater role in world markets and therefore gone outside the planned context of the domestic economy, some investment has had to be based on anticipation rather than demonstrated or firmly planned demand. Examples are ocean shipping and the Baykal-Amur Mainline (BAM). There have been expensive gaps between anticipation and reality, as the latter case demonstrates. BAM was built partly to give Soviet Far Eastern natural resources access to markets around the Pacific Basin. Unfortunately other suppliers competed strongly, the world economy went into recession, commodity prices dropped and so far the market for raw materials has proved much smaller than expected – a problem not confined to the Soviet Union.

The various factors which have frustrated Soviet efforts to minimize investment, as well as the shifts in policy during the past three decades and the growing spatial separation of natural resources and markets, have maintained transport's share of total Soviet investment despite the demands of other sectors in an economy trying to diversify. That is not to say that investment in transport has become generous. With the possible exception of inland-water transport, no mode appears to have any surplus capacity to speak of – though lack of capacity can have other causes than inadequate investment, as the next section will demonstrate.

Influences on demand

The preceding paragraphs suggest that investment policy and practice alone are sufficient to keep most Soviet transport operating close to capacity. However, there is another set of forces that tends to reinforce the pressure on capacity and therefore the pressure to invest more, by maximising demand. They differ from the forces that most often create high demand in capitalist economies. There, it often reflects a users' preference for versatility and convenience over immediate cost minimization, most frequently where road transport is concerned. It may also reflect politically-motivated subsidization or preferential regulation, as in the case of some shipping and air traffic, urban passenger transport and much road traffic, the last in response to road-users' lobbies. In the Soviet Union the consumers of transport services may have similar preferences and desires, but the latent demand has usually been suppressed by a balance of political forces different from that in most capitalist countries. In particular the road-users' lobby is weak and in general planners' preferences have outweighed those of consumers. However, such demand-reducing characteristics are offset by demand-enhancing features of the way transport is organized.

Firstly, transport like other sectors of the Soviet economy has traditionally used non-monetary indices of performance. Satisfactory performance has consisted in maximizing such indices as ton-kilometres or tons-carried, just as satisfactory performance in factories has consisted in maximizing tons produced. Secondly, where money has been used to calculate profits in a non-market economy, the profits have often been calculated as a mark-up over delivered costs. These two factors together have given both transport enterprises and their customers an interest in maximizing transport, sometimes even to the extent of sending

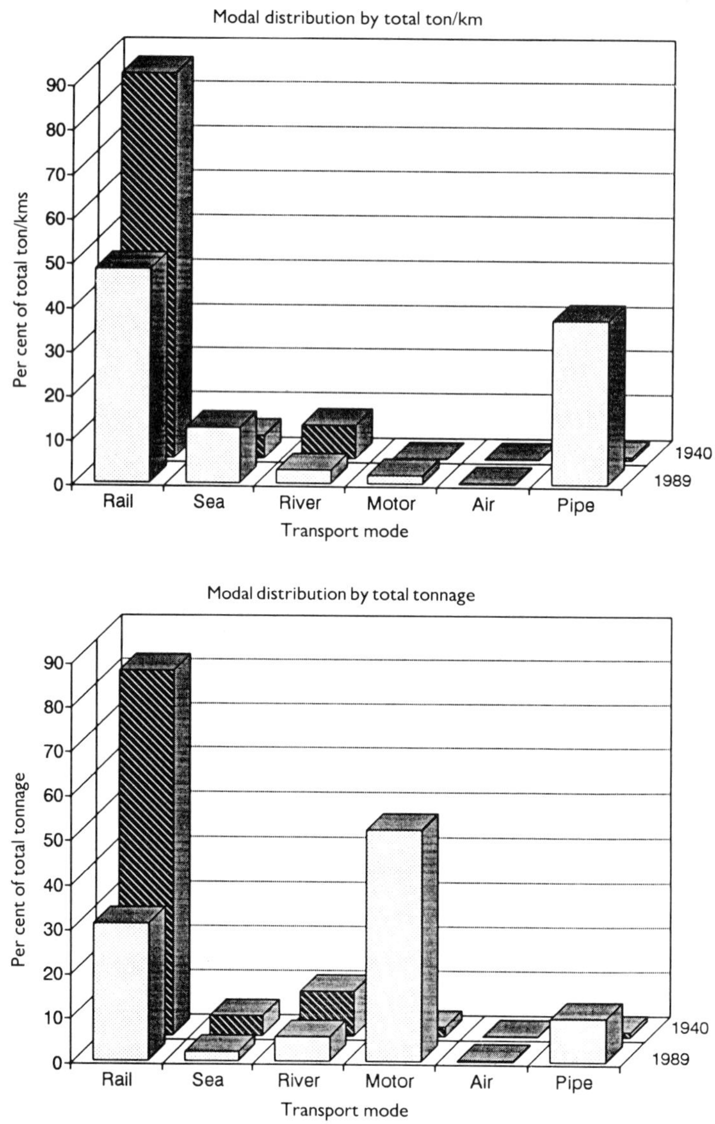

Figure 8.1 *The structure of common-carrier transport in the USSR, 1940 and 1989*

Table 8.1 *Railways' share of total national investment, 1918–88*

1918–28	9.0	1956–60	3.4
1928–32	10.2	1961–5	3.3
1933–7	10.6	1966–70	2.8
1938–40	10.7	1971–5	2.7
1941–6	11.6	1976–80	2.8
1946–50	7.7	1981–5	2.9
1951–55	5.0	1986–8	2.6

Sources: Narodnoye khozyaystov SSSR v 1970 godu (1971) Finansy i statistika, Moscow, pp. 482–3; *Transport i svyaz' SSSR* (1990) Finansy i statistika, Moscow, p. 134.

goods by roundabout routes.

Other factors, while not acting directly to elevate demand, reduce the incentive to control transport volumes and costs. For example, the Soviet railway ministry has in the past contracted with some production ministries to meet all their transport needs at a flat annual rate. Administrative costs are thereby reduced, but there is no incentive for the production ministries to minimize movements. Another example is the system whereby Soviet production enterprises sell not directly to consumers but to state-trading organizations. The production ministries are not motivated to take account of markets in locating production, and the trading organizations, acting as monopolies in a shortage-ridden economy, have no reason to pay close attention to transport volumes or costs.

In sum, transport in the Soviet Union has operated under pressures generated by policies that try to minimize investment and practices that maximize traffic. In consequence transport has acted as a brake on other sectors of the economy. Service has often been poor, and transport authorities, free from competition, have been able to pass on the costs of inadequacy to their customers. The railways, for example, have imposed high demurrage charges, refusing to guarantee arrival times for freight cars but charging heavily if they are not unloaded within a few hours of arrival. They have also moved goods according to priorities: high for coal and low for wood on the Trans-Siberian railway. With these systemic characteristics of traditional Soviet transport in mind, we turn to some specific problems faced by the system in the years immediately before perestroyka and the solutions attempted or proposed to counter them.

Soviet transport before Gorbachev: specific problems and solutions

Among the many problems facing Soviet transport by the early 1980s were some of particular interest to geographers, in so far as they concerned traffic flows, the evolution of the network and regional development. Three will serve to illustrate key factors in the transport situation.

The first problem was strain on the pipelines and, especially, the railways carrying fuels and raw materials from the eastern regions to the European Soviet Union. The problem reflected firstly the government's desire to maintain self-sufficiency in strategic raw materials; secondly the prodigal consumption of such commodities in heavy industry and the overwhelming dominance of that sector in the Soviet economy; thirdly the exhaustion of resources in the western Soviet Union, which contains three quarters of the country's population and industry; fourthly the need to supply socialist eastern Europe as well; and fifthly the lack of exports other than oil and natural gas that could earn hard currency. As all the sources of demand continued to grow, the search for new resources pushed further east and north and the strain on transport grew correspondingly (Table 8.2). Proposed solutions fell into two groups: those which recommended further investment in transport and those which sought to avoid it. The former included building more pipelines; expanding the capacity of existing railways; building new high-capacity railways; and adopting or inventing new transport-technology altogether, such as coal slurry or capsule pipelines or ultra-long-distance electrical transmission. The latter included switching from coal and oil to nuclear power in the European Soviet Union; reducing raw-material and energy consumption per unit of output in existing industries (for which there was great scope, to judge from a comparison of Soviet with Japanese or Swedish figures); altering the industrial structure of the European Soviet Union and concentrating energy-intensive industries east of the Urals; reducing the flow of supplies to eastern Europe; and finding alternative sources of foreign currency.

Given time and adequate funds for investment, some combination of these measures could have mitigated the problem. Unfortunately, under steadily deteriorating economic conditions, neither time nor adequate funds were available and the situation was further compromised by the incautious use of funds that did become available. One case has been mentioned already: the BAM represented in part an effort to open up new hard-currency markets in the east and relieve a component of the strain on routes westwards. Another case is the chain of nuclear power stations from the Baltic to the Black Sea, designed to supply both the western Soviet Union and parts of Eastern Europe, which included Chernobyl. A deeper problem was that the efforts to improve efficiency and reduce demand were counteracted throughout industry and transport by the systemic tendency to maximize demand at the expense of efficiency. For example, just as transport organizations could gain advantage in the tussle for funds, supplies and personnel by demonstrating an overwhelming growth of traffic, so production ministries could show irrefutable

Table 8.2 *Average distances carried per ton of freight by common-carrier transport, 1940–88*

	1940	1960	1970	1980	1989
All transport including:	660	452	493	543	630
Rail	695	798	861	923	962
Pipeline	481	422	858	1910	2357
Sea	758	1732	4051	3716	4040
Inland waterway	489	474	486	431	346
Motor	17	16	17	20	21
Air	397	908	1369	1716	1941

Sources: Transport i svyaz' SSSR (1990) Finansy i statistika, Moscow, p. 44.; *Narodnoye Khozyaystvo v SSSR 1989g* (1990) Finansy i statistika, Moscow, p. 559.

need by having many new projects, all delayed in construction by lack of supplies.

A second and related problem was that of transport for the North. It was related in that the demand for transport stemmed mainly from the search for energy and raw materials, and in that forces both within and outside the transport industry tended continually to maximize that demand. An extra element in the situation, however, was that the efficient provision of transport for the North required close intermodal cooperation. This did not apply so much to the shipping-out of raw materials: oil and gas was almost invariably moved out by pipeline and gold and tin by airn, for example. But it did apply to the provision of supplies. The lifeline of much of the North for this purpose was river transport. River transport could move freight cheaply but its navigation season was unpredictable. Goods for delivery by a deadline would therefore be entrusted to it only if a complementary mode, probably motor transport, was ready to take over at freeze-up or low water. Similarly, efficient service for the north-east Arctic coast required flexible cooperation between river and sea transport, since the overlap between river, coastal and deep-sea navigation seasons varies from year to year. Once again, efficiency was inhibited by systemic shortcomings: intermodal cooperation, though vital for the river-transport authorities, ranked low for the maritime ministry and the railways. Thus the maritime and river authorities developed incompatible equipment for use along the north-east coast, such as barges designed for an ocean-going barge-carrier that were unsuitable for use on the northern rivers. The railways, conscious that they were required to support both concessionary tariffs for traffic to the eastern regions and the expensive operation of northern 'pioneer' railways through cross-subsidy, were neither enthusiastic about facilitating traffic to river ports serving the North, nor about extending their own network northwards.

The solutions attempted and proposed for northern transport combined investment and reorganization. For example, riverport and seaport intermodal committees were set up to improve cooperation, and funds were provided to improve facilities. As with most such efforts at reform, there was immediate improvement in specific cases, but the systemic problems remained, limiting the extent and duration of improvements.

The third problem facing transport was that of supplying the countryside and moving agricultural produce out – a problem mainly of roadlessness. The extra element in this situation was long-term neglect. In the cases of east-west and northern transport much investment took place, but its effects were offset by traffic-maximizing tendencies. In the case of rural transport, relatively little had been invested. The problem was partly of systemic origin, reflecting the weakness of the road-users' lobby throughout Soviet times, and partly a result of a conscious policy to assign surface-transport funds to the railways. Solutions have focused on regional roadbuilding programs, for example in the non-chernozem zone of the European Soviet Union, but one wonders how effective this can be in the long-term without a reorganization of road transport to provide both efficient service to the rural communities and long-term lobbying power for investments equivalent to that wielded by the railways.

This section on the pre-perestroyka situation has tried to show how some of the specific problems facing transport remained without solution partly because of policy decisions by the central government, but principally because of systemic shortcomings common to the whole economy and well-exemplified by transport. It was these systemic shortcomings that perestroyka was intended to tackle.

The intentions of perestroyka

Intentions for the whole economy

Perestroyka was intended to make the whole economy more efficient. Three specific intentions are particularly relevant to transport. The first was to redirect investment to where it could better promote economic growth and development. Changes in either the sectoral or regional balance of investment would inevitably alter the national and regional demand for transport. The second intention was to increase the incentives for organizations and individuals to work better. Since 'better' meant not just 'harder' but also 'to better purpose', it implied new criteria for successful performance. A shift away from the maximization of physical quantities, such as output, goods handled or sales as a measure of performance was obviously to be expected. In the initial stages it would probably reduce the demand for transport but not evenly across the country.

The third intention, without which the other two could hardly succeed, was to reform the pricing system. Under the established order, thousands of prices were set by the central government, many based on social rather than economic criteria. Only if they more accurately reflected the costs of production and, principally, the demand for products would it be possible to make reasonable estimates of the relative utility of different investments and of what constituted effective work, and to shift from administrative to economic methods for guiding the economy. The then Prime Minister Ryzhkov, in his speech at the end of May 1990, summed up the situation when he stated that current prices in the Soviet Union reflected neither costs of production, nor supply-demand relationships, nor the level of world prices; that they could not even be used to analyse the economy; and that the experience of enterprises that had attempted financial and organizational reform had shown that it was the price and tariff system that was holding up economic change (*Izvestiya*, 25 May 1990, p. 2).

Price reform could bring substantial changes in traffic flows. For example, a persistent distortion in the established system has been the underpricing of raw materials and their transport relative to finished products. However, to forecast the impact of reform with any precision one would need to know details of cushioning subsidies, which the government would presumably provide, and whether price reform would be accompanied by the internationalization of the economy. The latter would involve switching to a convertible currency and opening up the domestic market to foreign competition. In the extreme case of all three measures being implemented without cushioning subsidies, many factories, mines and lumber camps would close, especially in the east and north. Flows of manufactures and basic raw materials could change radically.

The Soviet government has been unable to pluck up the courage to undertake sudden and complete price reform. Instead, according to the 'Guidelines for Stabilizing the Economy' approved by the USSR Supreme Soviet in October 1990, prices are to be freed over a period of two years or more, beginning with a group of non-essentials in 1991 (*Izvestiya*, 27 October 1990, pp. 2-5). Even then the freedom will be conditional. Republic and even local authorities will be able to freeze prices at will, which should make it easy for them to frustrate private entrepreneurship if they so wish. In the meantime, efforts are being made to revive the economy with the flawed tools available. The central government is supposed to confine itself to strategic decision-making. Subordinate organizations, individuals and regions are being given – indeed in the first two cases have already had for up to four years – more autonomy and more financial responsibility, in an effort to turn them from order-takers to initiative-takers. Formerly, for example, it was common for all revenues to be gathered into the central ministries and all expenditures disbursed from there on application. Now, subordinate organizations are being forced by 'full cost-accounting and self-financing', to use the favoured phrase, to make the connection between revenues and expenditures themselves.

Intentions for transport

Within the transport sector the main initial measure was to shift transport organizations at all levels, from the ministries down to individual ports and railway workshops, to full cost-accounting and self-financing. This meant that each organization was supposed to meet certain financial obligations to the state and to superior organizations; to provide for its own operating expenses and the maintenance and renewal of equipment; and to furnish social facilities and incentive payments for its employees. Revenues from the carriage and handling of freight and passengers were to be put into a set of funds for these purposes: a wages fund, a development fund, a social fund and so forth. These measures had the effect of devolving financial responsibility and some opportunities for initiative to the lower levels.

Full cost-accounting and self-financing in transport closely resembles that in manufacturing industry, for which it was designed. However, it was acknowledged from the outset that in transport the shift would be within limits set by the nature of the transport business. In the first place, self-financing would not include urban passenger transport, which would instead be required to achieve 'minimal losses'. This follows international practice in most big cities. In the second place, the degree of local autonomy implied by the new approaches may be achievable in automotive transport and to a lesser degree in water and air transport, but it seems particularly inappropriate for railways, which are still the principal middle- and long-distance carriers of industrial goods in the Soviet Union and therefore still the backbone of the country's transport system. Railways consist of hierarchical and interdependent units. Major investments especially must be coordinated: decisions on the type of electrification, loading gauge and even length of sidings cannot be left to local units. Soviet authorities think it unlikely that centralized investment on the railways will be cut below 60–70 per cent of the total. What is not clear is whether this centralized investment will be decided on the same bases as in the past. Investment decisions in other transport modes have already been decentralized, but not completely and not in the same way for each mode. Maritime and waterway transport, for example, both still have centralized ordering of vessels, but this is partly to achieve savings on multiple orders and avoid inter-company competition. In both cases individual companies have already chosen and bought equipment independently. River-transport companies will continue to have part of their route-maintenance costs covered by the central government budget, but maritime-shipping companies are now charged for icebreaking services. This is because many river-control projects are multi-purpose, whereas icebreaking costs relate to specific vessel movements (P'yanykh, Levitin and Kleyman, 1989, p. 24).

The autonomy of transport organizations is still restricted in other ways too. The government has not had the confidence to free the market completely for the carriage of goods. Instead, freight has been divided into two categories: *goszakaz* (literally 'state order') goods, which are basic industrial goods and must be carried at set tariffs, and contract goods, the tariffs for which can be decided, within limits, between the transport organization and its customers. However, there has already been some shift away from the *goszakaz* system. It originally applied to all goods for northern regions, for example, but in 1989 the only remaining state requirement was that the transport organizations involved be able to move a specified overall quantity of goods ('Towards the new', 1988, p .2).

The impact of perestroyka on transport

Problems of assessment

Perestroyka is a recent and current phenomenon. Its long-term effects on transport, if it continues, cannot yet be forecast with any confidence. This is especially so in the absence of so vital a measure as price reform. Also, most transport organizations only began full cost-accounting and self-financing in 1988. Some began in 1987 and others, in particular some union-republic road-transport ministries, in 1989. Two other factors hamper assessment of the role of perestroyka. First, its impacts overlap with those of earlier attempts at economic reform, including a tariff reform in 1982, a new pay system for road transport drivers in 1984 and new work indices in 1986. Second, the mechanics of implementing perestroyka in transport have not yet been perfected. Its introduction is in fact a classic example of the long-standing Soviet policy of 'leap before you look' or 'implement now, work out how later'.

Despite these limitations we can identify some initial impacts of perestroyka on transport. We can also detect a slowing-down of its impacts thereafter.

Transport operators' concerns and worries

The initial impact of perestroyka was to raise concerns and worries in the minds of transport operators. We shall look firstly at these concerns, then at the practical steps the operators took to cope with them.

The first major worry was that traffic would decline. An outsider might think that any reduction should be viewed positively, as a lessening of strain on an overburdened system. The concern of operators, however, was that the new financial system expected them to build up social and incentive funds and seemed to assume they would be able to do so because traffic would increase. The main reason for expecting a decline, especially in freight traffic, was that customers were expected to cut out the unnecessary and roundabout shipments that formerly served to raise the delivery price on which a profit mark-up was calculated, as well as sales that had always been unprofitable but contributed to plan fufillment. Some reductions in traffic did indeed occur quite soon – not everywhere, but sufficiently to confirm the transport operators' fears ('Na sovete otrasli', 1989, p. 2). There was also a source of traffic reduction within transport itself. Physical indices of performance were still used, but river and automobile transport switched from ton-kilometres to tons-loaded as a prime index, thus reducing the attraction of long-distance freight. In river transport the change was expected to reduce ton-kilometres by 1.3 per cent per annum for several years (Pronin and Sulaberidze, 1989, p. 6).

A second major worry concerned the failure of transport subunits to achieve real independence. They felt they had been made responsible for their financial fate without having control of it. This problem seemed to be worst on the railways and there were two aspects to it.

The first concerned the relations among neighbouring railway divisions, sections and stations. Nearly a year after the introduction of self-financing, the railways had still not worked out how to distribute the income from specific traffic movements among participating railways and subunits ('Nakanune partiynogo foruma', 1988, p. 5). To do so rationally would require three departures from previous practice. The first would be to separate out more precisely line-haul costs from the costs of loading and unloading freight. The second would be to replace the existing all-network average charges for loading and unloading, calculated for only ten broad categories of traffic, with charges better differentiated by region and type of traffic. The third change would be to work out a method for distributing line-haul income fairly among participating units when, as often happens, actual and tariff distances differ. An example of a practice that would have to be changed is that of allocating all income from the movement of refrigerated rolling stock to the railway owning it, despite the high costs of en-route maintenance. In the past this was of no consequence, since all revenues in fact went to the ministry and subordinate units were paid maintenance costs on application to the ministry. Now it is of consequence, since it discourages non-owning units from expediting refrigerated traffic.

The second aspect concerned relations among levels in the transport hierarchy, especially continuation of the overwhelming dominance by superior levels. Railway depots and shipyards soon began to complain they had been successful financially, only to see their profits syphoned-off to meet the deficits of their division or shipping company. Complaints became particularly sharp after the onset of civil unrest, since the railways especially demanded more from their subsidiaries to offset their losses in the troubled regions ('God ukhodyashchiy', 1989, p.3). The general problem was made worse by the fact that the railways were left by their ministry to decide for themselves how to arrange economic relationships with their subordinate units. There was therefore no national uniformity in such relationships. Some railways might require their divisions to hand over a percentage of profits; others demanded a set annual sum. The subordinate units could not protest effectively, since their legal status was uncertain. Resolution of the problem was hampered by a continuing dispute among economists and planners as to whether it was possible anyway to identify a separate product in transport below, say, the level of a railway division.

A third major worry concerned the performance indices used in transport. Government planners thought that a full shift to monetary indices, such as cost of movement, was not feasible in advance of national price reform. Physical indices therefore remained in use, though automobile and waterway transport, as we have seen, shifted from ton-kilometres to tons-loaded on the grounds that the former led to unnecessary movements in order to maximize bonuses. However, the new prime index can also be manipulated. It simply encourages transport companies to seek local rather than long-distance traffic. The railway ministry still distributes revenues to its subordinate units on a ton-kilometre basis (Avvakumov, 1989, p.64). Both of these indices, as well as such others in current use as average train weight and speed, have been criticized as not directly relevant to the improvement of transport's service to the economy. The concern of the operators is more that they are being motivated to structure their operations in a way that will have to be changed again when they are judged purely on financial grounds.

The fourth major worry was tariffs. As soon as transport organizations had to pay attention to revenues, they became concerned with tariff structures. Since then there has been a flood of complaints, but most fall into a few basic categories.

First, despite revisions (the latest in 1982), tariffs have failed to adjust to technical change. For example, when refrigerated freight cars replaced ice-cooled ones, tariffs remained the same. Customers benefited because spoilage was reduced but the railways faced higher running costs without compensation.

Second, tariffs are not finely categorized: they do not discriminate accurately among the various demands made on transport services. For instance, common carrier rates for road transport within union republics do not vary according to the quality of the road surface. Surface quality varies enormously. Also, it is assumed that medium trucks will carry at least six tons of freight. This makes the many movements in the 2-5 ton range expensive for customers, who therefore try to obtain their own trucks and avoid using the common-carrier fleet. Another case has already been referred to, namely the use by the railways of all-network average charges for loading and unloading. Yet another concerns icebreaker operations in the Arctic, especially the operation of expensive nuclear icebreakers by the Murmansk Shipping Company. From the 1950s to 1966 costs were borne by the state budget, which in effect made icebreaker service free to customers and costless for the operating company. Now tariffs are based on tons of cargo escorted. Since there are many empty return runs in the Arctic, the distribution of costs among customers must be quite haphazard. A final example concerns waterway transport. Despite tariff variations among river basins, the level of profitability varies greatly among shipping companies, types of traffic and types of operation. Overall profit levels on freight vary from 23 per cent for the Volgotanker company to 97 per cent for the Western company. On the River Lena, which has an average profit level of 49 per cent, the range is from 125 per cent for oil in tankers to 5 per cent for towing log booms (a disappearing activity) and 6.6 per cent for loading and unloading. One reason for the variations is that the current tariff structure tends to attract long-distance traffic whether it will bring profits or not. Another is that there are special low tariffs on routes parallel to railways, which have usually caused losses (P'yanykh, Levitin and Kleyman, p.24).

Third, and partly in consequence of poor categorization, transport organizations may be required to cross-subsidize within their operations. If road-transport rates do not vary with road quality, obviously movements on good roads are subsidizing those on poor roads. The principal mode affected, however, is railway transport.

Most railway traffic still moves at tariffs based on system-wide average costs. But there are special low tariffs for goods moving to the Far East, and the so-called pioneering lines in the North, especially the Baykal-Amur Mainline, are more expensive to operate than the average. Their extra costs are borne by the rest of the system ('Nakanune partiynogo foruma', 1988, p.3; Nikulin and Kovrigin, 1987, p.54; Kreynin and Mazo, 1987, pp.54–9). In the case of the line running north from Tobol'sk in Western Siberia, the subsidized rate drew traffic from river transport, which was much cheaper to operate, and caused congestion on the line.

Fourth, the problem of allocating revenues to the organizations incurring costs is not confined to the refrigerated movements mentioned earlier. For example, railway ferries are operated by steamship companies, not railways, and all revenues go to them. According to the railways, this means that their work in loading the ferries is unpaid. They have been lobbying to acquire control of the ferries ('Nakanune partiynogo foruma', 1988, p.3; Sekarev, 1989, p.30).

Shortcomings abound in the tariff system. They were tolerated in the past because the system had limited relevance. Now they are an urgent concern for organizations whose survival, or at least ability to make profits and retain good staff, may depend on them. This also means that the tariff structure, which used to be of little interest to geographers, is likely to have much more impact on the geography of traffic flows in the future.

The fifth and final worry to be examined here concerns equipment. In the past, transport organizations tended first to keep old or poor equipment in operation, since maintenance costs were not seen as an operator's problem; second, to obtain as much new equipment as possible, since over-equipment for the job was also not seen as an operator's problem, while inability to obtain equipment when needed was often a serious one; and third, to have every possible technical advance and option incorporated in new equipment, since versatility was considered much more important than ability to do a given job as economically as possible. They also often ignored depreciation, or allowed arbitrary and unrealistic rates set by superior organizations to go unchallenged. Perestroyka has brought many transport organizations to a state of alarm over their equipment. The Soviet merchant marine, for example, is one of the oldest in the world, with an average vessel age of fifteen years (Kornilov, 1988, p.8; 'Towards the new', 1988, p.2). Many vessels are too old to be allowed into the more discriminating foreign ports. Fleet renewal is hampered by a lack of foreign currency, coupled with the inadequate capacity and, for complex vessels, the inadequate expertise of Soviet shipyards. At a time of relatively high world prices, the position is especially bad for passenger vessels. In order to keep important domestic routes operating, especially in the Far East, it has been necessary to reduce Soviet participation in the lucrative international cruise market. This in turn has further reduced the merchant marine's foreign currency income.

There are many other examples of equipment worries in the merchant marine. The Kamchatka Shipping Company, which operates mainly old ships discarded by other companies, foresees bankruptcy if it is required to maintain or replace its fleet at its own expense. The Sakhalin Shipping Company has complained about a series of trouble-plagued ferries it has to operate between Sakhalin Island and the mainland. (The engines, designed for railway locomotives, were unsuccessfully adapted for marine use and the hulls have been corroding unusually fast.) Shipping companies working in the Arctic now describe the SA–15 icebreaking freighter and the KA–32c ice-reconnaissance helicopter, once lauded for their technical sophistication, as over-complicated and excessively costly to operate. The SA–15 is a particular irritant. In addition to having many features which are never or rarely used, it is extremely expensive to operate outside the Arctic. One major user, the Far Eastern Shipping Company, has to have 35–40 per cent of its fleet with icebreaking capability, but for much of the year the ships operate in open water in competition with ordinary freighters. Finally, river-shipping companies that use river-sea vessels to conduct foreign trade are concerned about the cost of replacing them. Most of those vessels were bought abroad. Prices have since risen more steeply than was expected when depreciation rates were set.

Positive reactions by transport operators

Initial reactions to perestroyka varied among transport operators. Most perforce reflected short-term horizons. From the point of view of impact on the transport process, a broad distinction can be drawn between positive and negative reactions, and within each category between conservative, entrepreneurial, manipulative and agitatory reactions. Operators' positive reactions tended to be conservative, entrepreneurial or agitatory.

On the conservative side, they began to cut out sources of waste and unprofitable activities. They sold off surplus assets, cancelled orders for new equipment, reduced manning levels and cut out unprofitable traffic. Announcements of surplus equipment for sale, rarely seen in the old days of hoarding, started to appear regularly in the technical-economic journals. Several river-shipping companies put vessels on sale, in some cases before taking delivery, and the operators of the East Germany-USSR (Mukran-Klaypeda) train ferry in the Baltic indicated that they would probably not want all the ferries on order. River-shipping companies eliminated some of their foreign traffic, not only because of the difficulty of replacing old river vessels, but also because they had previously been carrying some freight at a loss in order to earn hard currency and boost their annual ton-kilometre totals. Long-distance traffic lost priority within the country, too, and several river-shipping companies switched much of their effort to their most profitable business, the dredging and delivery of sand and gravel over short distances (Berezkina and Kapustyan, 1989, p.6).

On the entrepreneurial side, transport operators began to offer better services to try to attract traffic. For example, the railways operating into Moscow offered home-luggage delivery, video-bars on trains and polywrapped bed linen – at twice the normal price ('Nakanune partiynogo foruma', 1988, p.4). Several river-shipping companies, rather than sell unprofitable passenger vessels, tried to arrange package deals with foreign-tour firms. And the Murmansk Maritime Shipping Company not only set out to interest world shipping in the escorted use of the Northern Sea Route, in order to recover some of the crippling cost of operating its nuclear icebreakers, but even organized an Arctic cruise for Western tourists on a nuclear icebreaker (*Vodnyy Transport*, 1990, No.4, p.12). In general, possibly because they were already accustomed to operating in a capitalist world, the maritime-shipping companies seem to have shown the most initiative, trying to attract more foreign traffic and entering into joint ventures with foreign partners.

The main positive reaction that could be called agitatory was an almost universal call for reform of tariffs and amortization charges (Nikulin and Kovrigin, 1987, p.54). The transport companies especially wanted to eliminate cross-subsidization as a way of providing service to pioneer regions, and to differentiate charges by region and quality of service. Those transport companies that earned hard currency also lobbied to be able to use more of it for their own purposes.

The fading impact of perestroyka

Transport performed poorly in 1989 and 1990 ('God ukhodyashchiy', 1989, pp.2-4; 'Doklad gosudarstvennogo', 1990, pp.1-3). There were several reasons, some related to perestroyka. In the first place, not all the initiatives taken by transport organizations had as positive an impact as those described above. Some indeed probably reduced the likelihood of further progress in the sector, as a few examples will illustrate.

First, the fear of losing money led transport organizations to withdraw from unprofitable business (which had previously been attractive because it boosted the old performance indices) rather than try to improve the way they conducted it. Examples of withdrawal or attempted withdrawal included maritime shipping to Cuba, Vietnam and India, some operations in the Baltic by river transport companies using their river-sea vessels, and several multimodal operations. The Trans-Siberian Container Landbridge, an important generator of hard-currency income, lost business partly on account of poor cooperation between rail, sea and motor transport. Since multimodal operations were traditionally seen by transport organizations as peripheral to their area of interest, they were the first to suffer in times of stress (Gerasimchuk, 1990, pp.2-3; Miloslavskaya,

1988, pp.39-41; 'My vse', 1989, p.4).

Second, the lack of confidence in transport as a generator of income led to a diversion of entrepreneurial effort outside transport. Railway workshops took to making doors, windows and garden furniture, and ship-repair yards to producing prefabricated garages. This kind of activity was encouraged by the ministries: the problem was excessive enthusiasm ('Na sovete otrasli', 1989, p.5). In the transport press, therefore, there first appeared glowing reports from the local level of substantial improvements in income. Then there followed accounts from the national level of the shocking state of repair of much equipment and of huge backlogs at repair facilities (Travnikov, 1989, p.5; 'Ekzamenuyet zima', 1989, pp.2-4; 'My vse', 1989, p.3; Berezkina and Kapustyan, 1989, p.7).

Third, an approach of the manipulative variety was to reject the plan targets suggested, but no longer imposed, from above and restructure transport work for greater profit. In particular, the unilateral reduction of *goszakaz* quotas could leave room for more profitable contract work, including if possible the earning of hard currency (Berezkina and Kapustyan, 1989, p.6; Pronin and Sulaberidze, 1989, p.6; 'Po puti radikal'nykh', 1989, p.2). Unfortunately ambition sometimes outstripped capabilities, with disastrous consequences for other transport operators. Some ports, for example, accepted too much contract freight for export. Warehouses and approach railways became clogged, causing the railways and shippers to lose money though the ports made good profits ('My vse', 1989, p.4; *Morskoy transport*, 1989, p.2).

Fourth, realizing they were operating in an economy of shortages, some organizations began to gouge on prices or to impose side conditions in dealing with other transport organizations. Some ship-repair yards became so avaricious that their customers turned to foreign yards in desperation, using up the foreign currency intended for buying new ships ('Morskoy transport', 1989, p.2; 'My vse', 1989, p.4).

Despite their deleterious impact on other transport organizations, all these initiatives reflected a clear awareness of economic realities and possibilities at the local level. For the most part they reflected a lack of confidence in transport as a generator of income under existing financial and administrative conditions. Many transport organizations managed to raise their incomes substantially without any improvement in the standard measures of productivity or output.

A second reason for transport's poor performance by 1990 was that despite local successes in the short-term, and even if one ignores the problems attendant on those successes, longer-term prospects for transport were not good under the new conditions. Several problems emerged after a year or so. First, shortages of both consumers' and producers' goods meant that higher wages did not necessarily translate into better living standards, nor surpluses into new equipment. This reduced the incentives for both workers and managers. Second, much transport equipment was in need of replacement. Although railways and shipping companies could often find enough money to boost wages or build more housing for their workers, few could see how they were going to raise enough to replace major equipment. To make matters worse, much of what they needed would have to be bought abroad, given the lack of capacity and, often, expertise in the Soviet Union. Not only would that mean finding foreign currency at a time when the rouble was losing value, but also neglect of depreciation in the past meant that the necessary reserves had not been accumulated in any currency. Failure to replace equipment meant facing an ever-mounting bill for maintenance.

Third, under existing prices and tariffs it was not always possible to break-even at any level of efficiency. The deficiencies of the tariff system have already been described. Suffice it to add in illustration of their overall impact that seven river-shipping companies (the level immediately below the ministries) cannot break-even under the present system (P'yanykh, Levitin and Kleyman, 1989, p.24).

The performance of transport was made worse by a factor not directly related to perestroyka: the traumatic events of 1989 and 1990. Particularly disruptive were the ethnic strife in Transcaucasia and the railway blockade of Armenia; other service interruptions and bars to trans-border movements caused by ethnic tensions in Moldavia, the Baltic republics and Central Asia;

strikes in the major coalfields – the railways' biggest customers – and even in transport itself; and a series of disasters ranging from pipeline explosions to train crashes and the near-loss of the liner Maxim Gorkiy in the Arctic ('God ukhodyashchiy', 1989, p.2; *Izvestiya*, 28 September 1989, p.1). In addition to monetary loss and, in the case of the transport disasters, damage to morale, these events demonstrated that the Soviet transport system was in poor shape to cope with the unexpected.

However, probably the most important brake on further change has been the reassertion of the centralized administrative system. Two recent articles discuss this phenomenon in relation to the economy as a whole (Winiecki, 1990; Palei and Radzivanovich, 1990). In both the dominant theme is the resilience of the economic bureaucracy and Party apparatus, their capacity to absorb and nullify reforms that might threaten their dominance and in particular their ability to appropriate wealth from the system. Winiecki suggests they will especially resist decentralization and any expansion of the private sector. Either change will reduce the importance of their control of applications and permissions moving up and down the chain of command, with its potential kickbacks. It will also reduce the importance of the nomenklatura system and its associated political and economic networks. In other words Winiecki sees profit for the bureaucrats and apparatchiks stemming from control of process rather than economic growth. He describes several tested ways of thwarting reforms, including:

a) pseudo-reorganization. A few middle levels of the bureaucracy may be shed, but then new institutions are added to maintain the number of nomenklatura positions.
b) reducing the number of obligatory plan indicators for enterprises or shifting to new ones. The right choices can ensure problems and justify the return to more control.
c) abolishing commands but keeping control of supplies as a lever.
d) spreading propaganda against the reforms, claiming they will cause inflation or inequality.

Some current developments in transport can certainly be interpreted as fitting the model specified by Winiecki and implied by Palei and Radzivanovich. Recent lead articles in the technical-economic journals put out by, and presumably therefore purveying the views of, the transport ministries have focused on:

a) The anarchy consequent on subunits in the transport industry acting for themselves, especially in a shortages economy. There is, it is claimed, little incentive to cooperate, and little competition. Monopoly practices are the most likely outcome. 'They have their rights but are forgetting their duties,' in the words of the Minister of the Merchant Marine ('My vse', 1989, p.3). A basic problem, in the eyes of the ministries, is that the higher levels in the transport hierarchy have not yet learned how to replace administrative with economic levers and are therefore losing control of the transport process ('God ukhodyashchiy', 1989, pp.2-4).
b) The fact that the higher organizations retain responsibility for meeting targets and find it impossible to do so if they have no control (Gerasimchuk, 1990, pp.2–3).
c) The complexity of the process of moving to a new economic system. Revenues have to be allocated equitably among participants in the transport process. For that, and for price and tariff reform, much more information is needed, including information on practices in other countries. The ministries also need time to devise procedures to fulfil their new mandate of long-term planning. New cadres must be trained to operate the new system. All this leads inevitably to the argument that the demise of detailed central control can proceed in an orderly fashion only under detailed central control. Furthermore, it cannot be hurried ('Morskoy transport', 1989, pp.2–4).
d) Agreement on the part of the heads of subordinate units (e.g. the maritime shipping companies) that the central ministries should retain a strong role (Gerasimchuk, 1990, pp.2–3; 'My vse', 1989, p.4). This agreement probably reflects, first, a desire to retain a collective voice and lobbying agent in Moscow, as well as a unified commercial and legal policy *vis-à-vis* the foreign market. Second, it probably reflects a lack of confidence, stemming from a lack of experience, of the subordinate units in

their ability to operate in complete independence. However, it should be emphasized that this impression is given in the articles in the ministry journals. In personal conversation, shipping-company directors have stated that they scarcely need their ministry any more, although it remains useful in providing collective bargaining power when dealing with suppliers. Third, the subordinate units are probably influenced by the considerable power the ministries still wield over them – the power to appropriate their surpluses, for example – and by their own desire to reassert control over subordinate units.

The ministries' actions also fit the Winiecki model. At first glance they seem to have adhered to the spirit of reform. The Central Ministry staffs have been reduced, by some 35 per cent in the case of the Ministry of the Merchant Marine ('Morskoy transport', 1989, p.2). Many responsibilities have devolved to lower levels. Many norms and documents have been abolished and performance indicators have been replaced by simpler ones. A closer inspection, however, reveals measures to retain and reassert central control. Higher levels, as mentioned above, can appropriate the surpluses of lower levels and they retain some control over the allocation of resources. That some of the new performance indicators have proved to be no less susceptible to manipulation than their predecessors has helped to discredit them. And the maritime ministry, at least, has started to claw back the independence already granted to ports, ship-repair yards and the like. 'On the decision of the workers' these enterprises can retain their status as independent state enterprises or, instead, revert to structural units within shipping companies (Gerasimchuk, 1990, p.3). Since the ministry admits that workers are opposing reversion as inimical to their economic and social interests, there are presumably penalties for refusal ('Morskoy transport', 1989, p.2). It is certainly possible to justify the ministries' attitudes on the grounds that they are responsible for avoiding chaos during the transition to a new economic system. Clearly, too, many senior officials continue to believe in the old system, or at least in the possibility of reforming rather than abolishing it. Nevertheless the model of self-interest, specified by Winiecki and implied by Palei and Radzivanovich, is persuasive.

The economic reforms of 1990

The year 1990 was marked by protracted parliamentary debate over the next stage of economic reform. Proposals put forward by Prime Minister Ryzhkov at the end of May were sent back to be reworked after strong public objections. Rival groups of economists then proposed cautious and headlong approaches to reform; a compromise program was approved in October. The program has much more the flavour of caution and conservatism than of boldness and radicalism. There is little specific mention of transport but the implications are clear enough. Ministries and other senior administrative bodies will be required to supervise the diminution of their own power and status – precisely what is also envisaged by the transport ministries and a poor recipe for success. It is also exactly the opposite of what Palei and Radzivanovich regard as essential (Palei and Radzivanovich, 1990, pp.32–4). They believe that the only hope for reform is to put it in the hands of a special commission ranking above the ministries and supported by a dedicated research institute. Typical of the reforms, too, is the intended mixture of specified prices, prices regulated within limits and free prices, which should effectively stifle the impact of the last-named. In 1991 they are to comprise only 10-15 per cent of all retail prices anyway.

In sum, it appears that the 1990 reforms are more likely to confirm the conservative trends already emerging in transport than to provide a new impetus.

The long-term prospects for transport

The present political and economic uncertainty in the Soviet Union makes it impossible to assess prospects for transport with any confidence. However, at least three alternative national scenarios suggest themselves. The first two are the entrenchment of conservative trends and the rejuvenation of perestroyka, policy initiative remaining with the federal government in either case. The third is the growth of regional autonomy as an influence overriding either of the other two. The prospects for transport in each of these scenarios can themselves be disaggregated into prospects for investment, performance, route networks and traffic flows, and the division of traffic among modes. We shall attempt to forecast first national investment strategies if initiative remains with the central government, and their impact on transport; second the further progress and impact of perestroyka if it continues; and third the impact of regional devolution on transport. The potential impact of conservative trends in the central government seems to need little attention beyond references in the sections on investment strategies and perestroyka, if conservatism is assumed to imply reversion to a command economy. It would simply resurrect problems already discussed in the pre-perestroyka section of the chapter.

National investment strategies

If the centre retains the initiative in making major investments, some of the priorities and concerns expressed early in the Gorbachev era are likely to remain, whatever the fate of perestroyka. The re-equipping of industry west of the Urals will continue to be favoured, whereas remote-area investment schemes will be subject to more severe scrutiny than before Gorbachev's time (Bond, 1987, p.493). These generalizations should be tempered by the ambitious expectations for the Far East that emerged from Gorbachev's visit to Vladivostok in August 1986. Nevertheless it seems likely that investment projects for the Soviet North, at the very least, will be assessed more critically in future. The sectoral ministries evidently regard the eastern and northern regions as a bottomless pit for investment, producing low returns; demographers and regional authorities believe that far more people have moved to the North than are needed for the work to be done there; and, as mentioned earlier, the railways object to subsidizing unprofitable 'pioneer' lines.

In addition to overinvestment in the North for the work done there under present conditions, a further problem may arise if the Soviet Union achieves Gorbachev's twin goals of price reform and convertible currency. It may then emerge that some northern resource exploitation is uneconomic by world standards. Certainly the Soviet policies of strategic self-sufficiency in essential raw materials and low pricing of raw materials have produced more mining in the North than the Canadian approach of providing some government incentives to promote economic activity and then letting world markets decide what is attractive.

Oil and natural gas are so vital to the national economy, in an era of coalminers' strikes and public opposition to nuclear power, that they would seem the least likely resources to be affected by cutbacks. Admittedly lack of money has recently halted work on several northern gas pipelines; the investment program for the Yamal Peninsula is under review out of concern for the environment. In the longer term, however, more gradual, more economical and more environmentally sensitive development can be expected, probably with fewer people and smaller settlements, rather than any permanent rejection of the resources. The same may well be true of gold, tin and diamonds.

What then are the implications for transport? Gold, tin and diamonds need no bulk transport; oil and gas move out by pipeline. Therefore any reduction of northern economic activity, or of personnel and material requirements for current activities, would affect mainly the surface transport which brings in supplies and distributes northern resources for local use. In other words it would affect river and rail transport, Arctic shipping and, mainly on the local level, motor transport.

Transport will also be affected by the re-equipping of industry in the European Soviet Union and even more by any redistribution of industrial investment among branches of industry. Re-equipment should mean economies in the use of raw materials and fuel. They in turn could reduce pressure on the pipelines and railways from Siberia, offsetting in the latter case the growing share of the eastern regions in national coal supply. Redistribution of industrial investment away from basic heavy industries and armaments towards, say, electronics and biotechnology will have the same effect. It will probably also promote the growth of general and containerized freight transport as opposed to that of bulk raw materials.

Whether it will also, by requiring many small flows in many different directions, promote truck rather than rail transport, is a moot point. Having depended on rail transport so heavily in the past, Soviet industry is well located to continue using it more than other countries.

Perestroyka in other sectors of the economy

If present conservative trends continue, there may be little change in traffic flows beyond the loss of traffic already recorded as an initial impact of perestroyka, since the transport and production ministries will all want to preserve their spheres of activity. If perestroyka is rejuvenated, however, further substantial changes can be expected. The initial loss of traffic will presumably be offset by increases as the economy expands, if all goes well. Whether the new demands on transport will be similar to the old is another matter. For example, they might stem from shifts in the spatial distribution of manufacturing. Consider first the impact of ministerial self-financing, which has been in place for some time, and price reform, which began at the beginning of 1991. ('Price reform' here refers only to new, wholesale prices based on average industry-production costs and not to any move to market prices.) They will require change in the criteria for selecting new factory locations and may well force the closure of many factories with above-average production costs. Up to the present the latter have been kept open by the ministries' need to meet production quotas and their acceptance of cross-subsidization, provided that average branch costs are kept within acceptable limits. Cross-subsidization, however, is supposed to cease with price reform. The metallurgical industry claims that over a hundred of its factories will need subsidies to support their basic operations and over two hundred will require them to finance essential capital investments ('Metallurgical', 1990, p.10). One suspects that many of these factories, and similar ones belonging to other ministries, are in the eastern Soviet Union and will eventually close unless they can shift to locally-needed or higher-value output.

Secondly, consider the impact of making the rouble convertible. If the Soviet Union were to participate in international trade on a scale commensurate with the size of its economy, one would anticipate much more industrial activity in the country's ports, processing both imports and exports. As we have seen, even the limited expansion of trade already achieved during perestroyka has taxed the capacity of some ports.

The preceding paragraphs illustrate how the further evolution of perestroyka might affect industrial location and therefore transport. The range of possible impacts is much greater. For example, new but controlled wholesale prices are supposed eventually to give way to market prices. This will introduce yet another set of criteria for industrial location and survival. Coupled with the elimination of shortages, which could occur through either expanding output or opening the domestic market to foreign suppliers, it would make the less economic factories even more vulnerable.

Perestroyka within transport

Even if the emerging conservative trends continue, some of the improvements in performance described earlier might be maintained, especially at the local level. For example, maintenance and servicing at ports, railway stations and the like has been subcontracted to cooperatives with considerable success; there is no reason to suppose that this will not spread ('Obsluzhivaniye flota', 1990, pp. 2–3). Linking revenues with expenditures at all levels is a fundamental improvement, which can still benefit the industry if higher levels in the hierarchy can restrain themselves from interfering either by appropriating surpluses or by providing subsidies and cheap credit. The ability not to interfere depends in turn on the promised improvements to the tariff system. Also the whole concept of independent subunits is much more applicable to some modes of transport than to others – and least applicable to the railways. In almost all countries they have traditionally relied on strong discipline in a closely-coordinated hierarchy.

The railways view the future with particular foreboding. It is intended that they will remain under All-Union ownership. Unfortunately the government has been slow to change the conditions under which they operate, to put them on even terms with customers and competitors operating in an emerging market economy. They complain of excessively low tariffs, rigidly regulated; of a burgeoning deficit that hampers technical renewal and the provision of employee benefits; and of a lack of cooperation from regional authorities that have no proprietary interest in them ('We request help', 1990, p.10).

For any mode of transport, a strong incentive to efficient performance will be the opportunity to conduct business in foreign markets and earn hard currency. This favours first maritime shipping, with its long experience in the field and record of initiatives already under perestroyka (Brown-Humes, 1989, pp.6, 9, 14); second the river-sea operations of the river-shipping companies; and third the air transport industry. Efficiency in conducting foreign business is not necessarily accompanied by efficiency in meeting domestic needs.

The desire of several transport organizations to cease handling unprofitable traffic could drastically affect traffic flows if it were allowed free rein. That is unlikely if the traffic has to be carried in the national interest. The central government will either include the neglected traffic in *goszakaz* allocations or, eventually, adjust tariffs to make it profitable for someone.

If conservative trends persist, the continuing dominance of the transport ministries will make unlikely what could have been the most significant change in transport to emerge from perestroyka: a major shift in the relationships among transport modes. This would have occurred through the formation of true multimodal companies after the fashion of Canadian Pacific or TNT in Western countries. Multimodal organizations have been proposed in the Soviet Union, mainly for ports and occasionally for specific traffic flows, but something more comprehensive is needed. Western multimodal companies carry much of the general and fast traffic that does not move efficiently in the Soviet Union, but is likely to become increasingly important if the industrial structure of the country changes as the government intends. Control of two or more modes gives such companies speed and flexibility.

A bonus of the emergence of multimodal operators might be to jerk Soviet motor transport out of its lethargy. Much of the country's motor transport consists of subsidiary operations run by production ministries and their enterprises, for whom it is not a priority to keep abreast of the latest developments in the field. Common carrier-transport has a reputation for poor reliability and inflexible tariffs. One of its unpromising reactions to perestroyka has been to plead for a stronger monopoly. In sum, road transport has not been a source of constructive initiatives for change. This is in complete contrast to Western and many developing countries, where road transport has provided innumerable opportunities for private entrepreneurs. If perestroyka can bring comparable scope for initiative in the Soviet Union, then for transport at least it will indeed have brought about the revolution referred to at the beginning of this chapter.

The growth of regional autonomy

The final scenario suggested above would involve the disruption of both perestroyka as originally conceived and the emergent conservative trends. In fact the seizure of initiative by regional authorities has already disrupted transport and traffic flows. In 1990 Azerbaijan cut off railway traffic to Armenia. Also in 1990, in response to serious food shortages, authorities in the RSFSR raised the procurement prices for agricultural products. This led to privately-initiated exports from the Ukraine, which were eventually prohibited by the republican government. Moves towards independence in the Baltic republics have had a more serious impact. The federal government tried to bring Lithuania into line by cutting off supplies of oil and natural gas; Lithuania responded by setting up customs posts and reducing food exports. In addition to all this, most republics have initiated bilateral trade agreements with other republics, bypassing the federal government.

Quite apart from what might be forced on the federal government, greater economic autonomy is to be granted to republics and to regions of the RSFSR. The exact terms and conditions will depend on the final form of the new All-Union treaty, assuming that a version can be devised that most republics can be persuaded to sign. Regional autonomy in whatever form has the potential to bring great changes in transport. The major interregional flows of raw and semi-processed materials, and the investments to facilitate them, are unlikely to change in the near future. In other respects, however, investment will probably be reoriented to meet regional and local needs, in order to counteract the overwhelming priority given to national requirements in the past. This implies attention to local air services in the larger regions; possibly river transport, since many river-shipping companies are essentially regional in their orientation; and above all regional and local roads, so much neglected in the past. Also local enterprises may well start to seek more of their inputs locally rather than through their ministry networks. It may be a sign of things to come that during the federal embargo on fuel deliveries, Lithuania set about opening up some small oil wells that had been capped because their flow was inadequate to interest the federal ministry. Greater interest in local sources of supply should in turn stimulate local road transport – including and especially cooperatives.

In addition to all this, signs are already appearing of action on a wider-than-local scale. Latvia, for example, has taken substantial initiatives in maritime shipping and the incorporation of a regional airline. In Yakutia, to judge by comments during interviews in the summer of 1989, regional decision-making could well mean not only increased investment in road transport within the republic but also abandonment of the AYAM railway project, which would have provided Yakutia with its major external link. It could also mean the introduction of air services catering to tourists, in cooperation with the new Latvian airline rather than Aeroflot. Alaska Airlines has already contracted to provide new services from the Soviet north-east to the USA. In the longer term, peripheral republics might well replace the RSFSR with foreign suppliers for some commodities and in general trade more outside than within the Soviet Union. In sum, regional autonomy, an outgrowth primarily of glasnost, will probably stimulate far more radical change in transport than all the efforts of perestroyka.

There is, however, one factor that makes it hard to forecast the impact of regional autonomy with any precision. That is uncertainty as to the type of economic policy autonomous regions will follow. On the one hand it has become almost an article of faith that the best type of economy is a market economy. On the other the initial actions of regions that have already seized the initiative suggest substantial government interference in the market can be expected. This is not surprising. In the first place the habits of a command economy are not easily discarded. In the second the peripheral republics are small relative to the RSFSR. They will not want to be dominated economically and they are bound to be nervous of the intricate linkages that currently bind them together. Perhaps in a few years we shall see among them the range of policies now found among developing countries trying to industrialize, from import substitution behind tariff walls at one end of the spectrum to export-led development at the other. Alternatively there might emerge a Soviet common market after the fashion of the EEC. Obviously the implications for transport would be very different in the two cases.

Conclusions

Perestroyka seems to have had a more significant initial impact than many earlier reforms, both on the Soviet economy in general and on transport in particular. During 1989 and 1990, however, it appeared to be reaching or to have reached the limit of its impact in transport. A major reason was that, as in the case of all previous reforms, established interests and viewpoints were proving too strong. However, a new force had already made itself felt and seemed likely to have a much greater impact in the long-term. In most parts of the country the drive for regional autonomy has far greater appeal both to the bulk of the populace and to local authorities and politicians, despite disagreements over how far autonomy should be taken. It is not seen as an initiative from the Centre; it appeals to the heart as well as to the mind; and it opens up previously unimagined possibilities for local control of the local destiny. For models of economic development, and therefore the evolution of transport, under such an impetus we should perhaps look not to anything in the Soviet past, but rather to the post-colonial industrializing Third World, the EEC, or even indeed East-Central Europe after the First World War.

Suggested reading

Lydolph, P.E. (1990), 'Transportation and domestic trade', *Geography of the USSR*, Misty Valley, Elkhart Lake, Wisconsin, USA, pp. 371–404.

References

Avvakumov, S.M. (1989), 'Tarify, dokhody i khozraschet', *Zheleznodorozhnyy transport*, 12, p. 64.

Berezkina, G., Kapustyan, V. (1989), 'Perestroyka: real'naya ili formal'naya?', *Rechnoy transport*, 4, p. 6.

Bond, A.R. (1987), 'Spatial dimensions of Gorbachev's economic strategy', *Soviet Geography*, 28, 7, pp. 490–523.

Brown-Humes, C. (1989), 'Soviet Shipping: Special Report', *Lloyd's List*, 13 September, pp. 6–17.

'Doklad gosudarstvennogo komiteta SSSR po statistike', *Pravda*, 28 January 1990, pp. 1–3.

'Ekzamenuyet zima', *Zheleznodorozhnyy transport*, 1989, 11, pp. 2–4.

Gerasimchuk, G. (1990), 'Morskoy transport: problemy upravleniya', *Morskoy flot*, 2, pp. 2–3.

'God ukhodyashchiy: raschety i proschety', *Zheleznodorozhnyy transport*, 1989, 12, p. 3.

Kornilov, V. (1988), 'Perestroika in the Soviet merchant marine and liner shipping', *Soviet Shipping*, 3, pp. 8–9.

Kreynin, A.V., Mazo, L.A. (1987), 'Tarify v usloviyakh samo-finansirovaniya i samookupayemosti', *Zheleznodorozhnyy transport*, 6, pp. 54–9.

'Metallurgical industry sounds the alarm', *Current Digest of the Soviet Press*, 42, 38, p. 10. From *Pravda*, 24 September 1990, 2.

Miloslavskaya, S.V. (1988), Smeshannyye perevozki v novykh usloviyakh khozyaystvovaniya, *Zheleznodorozhnyy transport*, 2, pp. 39–41.

'Morskoy transport: perestroyka i upravleniye', *Morskoy flot*, 1989, 5, pp. 2–4.

'My vse – doverennyye litsa flota', *Morskoy flot*, 1989, 12, pp. 2–5.

'Nakanune partiynogo foruma', *Zheleznodorozhnyy transport*, 1988, 6, p. 5.

'Na sovete otrasli', *Rechnoy transport*, 1989, 10, pp. 2–5.

Nikulin, S.A., Kovrigin, A.G. (1987), 'Na puti k polnomu khoz-raschetu', *Zheleznodorozhnyy transport*, 8, pp. 52–5.

'Obsluzhivaniye flota – na khozraschet', *Morskoy flot*, 1990, 3, pp. 2–3.

Palei, L.V., Radzivanovich, K.L. (1990), 'How to carry out economic reform: points of view and reality', *Soviet Studies*, 42, 1, pp. 25–38.

'Po puti radikal'nykh preobrazovaniy', *Zheleznodorozhnyy transport*, 1989, 3, pp. 2–6.

Pronin, V., Sulaberidze, V. (1989), 'Nuzhen sleduyushchiy shag', *Rechnoy transport*, 10, p. 6.

P'yanykh, S., Levitin, V., Kleyman, Z. (1989), 'Novyye tarify', *Rechnoy transport*, 2, pp. 24–5.

Sekarev, Ye.I. (1989), 'Paromnyye kompleksi: ekonomika vsaimo-deystviya', *Zheleznodorozhnyy transport*, 1, p. 30.

'Towards the new economic system', *Soviet Shipping*, 1988, 4, p. 2.

Travnikov, A. (1989), 'Delo gosudarstvennoy vazhnosti', *Morskoy flot*, 5, p. 5.

Vodnyy Transport, (Water Transport, Journal).

'We request help', *Current Digest of the Soviet Press*, 42, 38, p. 10. From *Pravda*, 17 September 1990, p. 2.

Winiecki, J. (1990), 'Why economic reforms fail in the Soviet system – a property rights-based approach', *Economic Inquiry*, 28, 2, pp. 195–221.

9

Foreign trade and Soviet regional development

Michael J. Bradshaw[1]

McDonald's
McDonald's

Introduction

In the early 1990s Western governments are torn between providing economic aid to support reform in the Soviet Union and withholding aid to protest about the actions of the central government in Moscow. For many observers the opening of McDonald's on Moscow's Tverskaya Street (formerly Gorky Street) is symbolic of the increasing internationalization of the Soviet economy. Increased foreign-trade activity is seen as a vital component in the transition of the Soviet economy to a market-based system. The Soviet Union is making an unparalleled effort to join the world economic system and is seeking membership to international institutions such as the International Monetary Fund, GATT and OPEC.

This chapter examines the recent reforms of the foreign-trade system and assesses how these changes are influencing and influenced by the relationship between foreign trade and regional economic development. As with so many aspects of Soviet economic reform, it is necessary to start by examining the system prior to 1985. The first half of this chapter considers how the centralized control of foreign-trade activities influenced industrial and regional development in the Soviet Union. The initial discussion considers trade with all partners, but the more detailed analysis that follows concentrates on trade with the industrialized West. Since the dramatic events of 1989, Soviet trade with the nations of Central Europe (previously the Soviet Union's socialist partners) is in a state of flux as transactions are transferred to convertible currencies. In the final analysis it is likely that Soviet trade with Central Europe will be managed on a similar basis as trade with the West; focusing on East-West trade therefore provides a more general insight into the future organization and impact of foreign trade within the Soviet economy.

Foreign trade under Soviet central planning

Because of the size of the domestic economy, substantial natural-resource wealth and a long-standing policy of economic self-sufficiency, one would not expect the Soviet Union to be a major player in world trade. According to the International Monetary Fund, the Soviet Union's share of world exports in 1970 was 4.2 per cent, and its share of imports was 3.7 per cent; the value of total trade turnover was $ 24.5 billion. In 1985 the Soviet share had risen to 4.7 per cent of world exports and 3.7 per cent of imports; the value of total turnover was $ 168.9 billion, with a trade surplus of $ 3.9 billion. By comparison, in 1985 Japan accounted for 9.5 per cent of world exports and 6.7 per cent of imports, while the value of total turnover was $ 307.7 billion and Japan's trade surplus was $ 46.7 billion. Not only has the Soviet Union been a modest actor in the world economy, it has also been a passive player in natural-resource markets. Although the Soviet Union is a major producer and exporter of many key mineral resources, it has seldom sought to influence world prices (Ballinger, 1985; Hewett, 1986). In many instances Soviet trade behaviour has been careful not to undermine resource prices by flooding markets and for the most part the Soviet Union has been a price-taker, not a setter.

Soviet hard-currency purchasing power benefited from the dramatic increases in energy prices in 1974 and 1979 and, conversely, was seriously damaged by the crash in oil prices in 1984. Any windfall gains from the Gulf War are likely to be limited because the Soviet Union is unable to expand production to profit from higher world prices. Although there has been a substantial increase in the value of Soviet foreign-trade turnover (according to Soviet foreign trade statistics the value of trade increased from 22 billion roubles in 1970 to 140.8 billion in 1989) the Soviet Union has not substantially increased its role in the international economy. The expansion of Soviet foreign trade is less a consequence of a substantial increase in the volume of trade than windfall profits from resource exports. Between 1970 and 1985 the value of trade turnover increased seven-fold, but the volume of trade barely doubled.

Because of the problems associated with the Soviet pricing system (some of which are discussed below) it is difficult to evaluate the importance of foreign trade to the Soviet economy. Shmelev and Popov (1990, pp. 222–3) suggest that in 1970 foreign trade accounted for 8 per cent of national income (4 per cent of GNP using Western methods); in the same year foreign trade accounted for 18 per cent of Japan's GNP and 8 per cent of the United States' GNP. By the mid-1980s the share of trade had increased to 25 per cent of Soviet national income (12–15 per cent of GNP) while the corresponding measure for the United States was 15 per cent of GNP. Analyses of the role of Western equipment in Soviet equipment investments (Hanson, 1981, p. 129; 1986, p. 61) have suggested that the share of Western imports as a percentage of total equipment investments peaked at 8 per cent in 1976, and during 1980-84, averaged 7.2 per cent. Clearly the increases in the value of Soviet foreign trade and its share of national income suggest that foreign trade has become an increasingly important factor in Soviet economic performance since the 1970s, but this aggregate pattern hides geographical and structural shifts that can only be identified by a more detailed examination of foreign trade under Soviet central planning.

The centralized management of Soviet foreign trade

Soviet foreign economic relations have been managed and controlled by the central planning system through the Ministry of Foreign Trade. This state monopoly of foreign trade dates from the late 1920s and remained essentially unchanged until 1987. Over this period a complex bureaucratic system developed to manage foreign economic relations (Gruzinov, 1975; Gardner, 1982). The Ministry of Foreign Trade served to integrate foreign-trade policy dictated by the Party and GOSPLAN with the operations of the industrial ministries and their associated enterprises. The Foreign Trade Organizations (FTOs), responsible for particular industries and subordinate to the Ministry of Foreign Trade, have acted as the link between the central ministry and the various sectors of the Soviet economy. The most important feature of the FTO system is that it has served to isolate Soviet industrial enterprises from the international economic environment. An industrial enterprise producing goods for export had little or no contact with the foreign partner, because the FTO handled the contract. At the same time the enterprise reaped scant reward from export production, providing little incentive to produce for export markets. Likewise the FTO handled the purchase of foreign machinery and equipment and foreign business established contacts with the relevant FTO rather than the final customer. Because of this isolation exporters were unable to react quickly to the needs of a foreign buyer and imported goods often failed to meet the needs of the recipient enterprise. As a consequence of the cumbersome bureaucratic system regulating trade, problems of corruption, poor integration and departmentalism have plagued foreign economic relations as much as they have the domestic economy.

Because the system of prices used in the Soviet economy has been based on different concepts and served a different purpose from prices in Western market economies (a distinction that market reforms aim to remove), it has been necessary to have separate external foreign-trade prices and internal prices. The situation is further complicated by the fact that the rouble is not a hard currency: it is hoped it will become convertible by 1995. When dealing on international markets Soviet traders have to use a hard currency, usually US dollars, which makes Soviet trade susceptible to fluctuations in the value of the dollar. In the case of intra-CMEA trade a convertible rouble has been used as an accounting unit to manage what is essentially barter trade between the Soviet Union and Eastern Europe. Since 1 January 1991 Soviet trade with the nations of Central Europe (who were members of CMEA) has been, in theory at least, conducted with hard currency.[2] Given that this is well in advance of the creation of a convertible rouble it can only further complicate foreign-trade transactions. Under the traditional foreign-trade system any good traded on foreign markets by the Soviet Union had at least three prices: first, the price paid or received for a good on the world market expressed in the relevant foreign currency; second, the same price converted into foreign-trade roubles on the basis of an official exchange rate (which has always been over-valued); and third, the domestic price paid to the producer of an export item or paid by the recipient of an imported good (in the most simple case the import price charged to a Soviet enterprise is calculated on the official exchange rate). A final complication is that there are about 10,000 'differentiated foreign exchange coefficients' used to calculate the domestic value of imports and exports (Shmelev and Popov, 1990, p. 229). Obviously this unwieldy management system has not only isolated Soviet enterprises from the world economy, but it has also made it impossible to assess the benefits of foreign-trade participation for a particular enterprise or region. In November 1990 the rouble was devalued for commercial purposes in an effort to introduce a single currency standard for Soviet enterprises, eliminating the need for a multitude of exchange coefficients.[3]

The geographical distribution of Soviet foreign trade

It should be no surprise that Soviet accounting practices, coupled with the impact of inflation, make it very difficult to measure and compare trade flows between the Soviet Union and the rest of the world. The foreign-trade handbooks published by the Ministry for Foreign Economic Relations in Moscow (now *Vneshniye ekonomicheskiye svyazi v SSSR*) divide the countries of the world into three main groups: Socialist countries (subdivided into CMEA and Non-CMEA), the industrially developed capitalist countries and the developing countries (Figure 9.1).[4] The classification used in Table 9.1 is somewhat different in that it uses major geographical regions. The distinction is made to examine geographical change in the pattern of trade flows. The increased share of trade with Western Europe is a consequence of the expansion of East-West trade during the 1970s; otherwise the pattern of trade is quite stable. Figure 9.1 shows the overall growth of Soviet foreign trade with all partners. These data hide a number of different trading relations. At the heart of the various trading relations are political and economic considerations. In general, trade with the other members of the CMEA has been based on a convertible rouble accounting system. Trade with the industrialized West is based on hard currency (the exception being Finland which until recently operated a barter system), while trade with the developing world is based on barter, often with the Soviet Union providing arms, a limited amount of hard currency and a considerable amount of foreign aid. Because of the different financial systems operating with each trade bloc it has not been possible in the past to use a trade surplus generated in Eastern Europe to purchase imports from Western Europe. It is, however, the case that the Soviet Union has used hard-currency surpluses generated in West Europe to subsidize trade with Japan and North America. In recent years Soviet-Japanese trade has declined, but South Korea is now emerging as a potentially important partner in the Asian-Pacific region. Unfortunately for the Soviet leadership and Western business, Soviet terms of trade have deteriorated in recent years. During the early 1980s the Soviet Union had a sizeable surplus in its trade with Western Europe, sufficient to cover the cost of grain imports from North America and imports of machinery and equipment from Japan. Similarly, the Soviet Union had a healthy surplus in trade with Eastern Europe which has now deteriorated into a sizeable deficit. By the end of 1990 the Soviet Union's aggregate debt to Eastern Europe stood at between 8 and 10 billion roubles.[5] In early 1991 the Soviet Minister of Finance declared the Soviet Union's total foreign debt to be 39 billion roubles ($ 62.4 billion). To understand this change in fortune it is necessary to examine the structure and dynamics of Soviet foreign trade.

The structure and dynamics of foreign trade

Soviet exports are dominated by trade in energy resources and raw materials, which have increased in importance since the early 1970s. The export of energy and raw materials is particularly important in trade with Eastern Europe and the industrialized West. According to OECD statistics, during the 1980–85 period exports of mineral fuels and energy accounted for an average of 78.4 per cent of Soviet exports to the West. One indication of the increasing role of oil and gas exports is the fact that in 1970 pipelines accounted for 12.7 per cent of export transportation; by 1988 their share had risen to 31 per cent. When one considers exports of cork, wood, diamonds and non-ferrous metals, the reliance on raw materials increases to

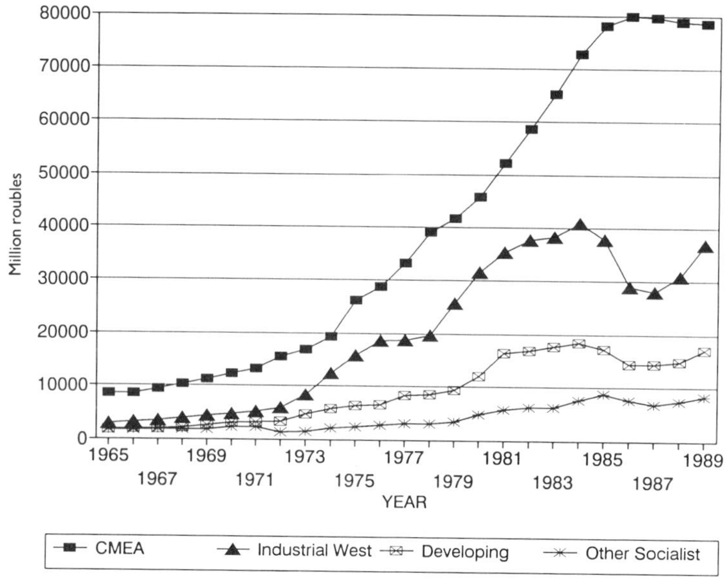

Figure 9.1 *Growth of Soviet foreign trade, 1965–89*

Table 9.1 *Geographical distribution of Soviet foreign trade*

	1970	1975	1980	1985	1989	1990[3]
Western Europe	16.6	24.4	27.0	21.3	20.5	22.3
Exports	15.1	24.4	29.7	23.8	21.0	24.5
Imports	18.2	24.4	24.0	18.8	20.0	20.4
Balance[1]	−187.5	−652.8	+4033.5	+4246.1	+11.4	+288.8
North America	1.3	4.1	2.7	2.6	3.9	3.0
Exports	0.6	0.7	0.4	0.5	0.8	0.8
Imports	2.1	7.1	5.2	4.8	4.5	4.8
Balance	−155.6	−1732.1	−2142.1	−2882.2	−2709.7	−2171.2
East Europe	56.9	48.9	46.6	51.5	51.0	48.0
Exports	55.3	52.6	46.3	50.6	50.0	45.0
Imports	58.6	45.3	47.0	52.5	53.0	50.6
Balance	+180.0	+561.0	+2113.2	+301.6	−4555.2	−6540.8
Asian-Pacific	7.6	7.3	7.2	7.2	8.8	8.6
Exports	8.1	6.2	5.3	6.8	9.3	9.5
Imports	7.1	8.3	9.2	7.6	8.3	6.3
Balance	+183.4	−718.8	−1461.4	−325.4	+381.8	+965.9
Other[2]	17.6	15.3	16.5	17.4	15.8	18.1
Exports	20.9	16.1	18.3	18.3	18.9	20.2
Imports	14.0	14.9	14.6	16.3	14.2	17.9
Balance	+183.4	−94.2	+2637.5	+1994.4	+3498.9	−420.8

Notes:

1. The balance is reported in millions of roubles. All other figures are percentage of total.
2. Includes the Indian sub-continent, Africa, Central and Latin America.
3. Because these data are based on a summary of foreign-trade returns there is a certain amount of unclassified trade activity included in 'other'.

Sources: Ministerstvo Vneshney Torgovli (Various Years), *Vneshnyaya torgovlya v SSSR*, Finansy i Statistika, Moscow. Data for 1990 are based on third-quarter foreign trade returns as reported in *Vneshnaya torgovlya*, 1990, (11).

an average of 86 per cent of exports to the West (Bradshaw, 1991). Machinery and equipment are more important in exports to the socialist and developing worlds. The reason for this is the inability of Soviet manufactured goods to complete successfully on Western markets.

The structure of Soviet imports is very different with the most important items being machinery, equipment and food products. The majority of Soviet machinery imports comes from other CMEA members. During the 1970s the industrialized West accounted for about 30 per cent of Soviet machinery imports; during the 1980s the West's share fell to about 23 per cent. Food products are imported from the major Western grain producers (United States, Canada, Argentina and the European Community) as well as from the developing nations. Since the late 1970s imports of food from the industrialized West have increased at the expense of machinery and equipment imports. During the 1980–85 period machinery, equipment and pipe imports accounted for an average 37.3 per cent of Soviet hard currency imports, while food imports accounted for 40.2 per cent of hard currency imports (Hanson, 1986, p. 60). Cole (1985, p. 373) has produced a classification of Soviet trading partners that links the geography of trade with its commodity composition. Three broad types of trading partner are identified: first, the countries of Europe (East and West) together with Japan, to which the Soviet Union supplies fuel and raw materials in return for manufactures; second, the developing countries to which the Soviet Union exports mainly manufactured goods, in return for agricultural products, raw materials and small amounts of oil; and third, the United States, Canada, Australia and Argentina, which supply the Soviet Union with agricultural products but import little from it. The dominant trade relationship is the export of energy and raw materials in return for manufactures.

Because of the non-convertible status of the rouble, which means that the Soviet Union must first export to hard-currency markets before it can purchase imports, and an excessive reliance on energy resources and raw materials, the dynamics of Soviet foreign trade reflect the changes in world-resource prices. As already noted, the dramatic expansion of Soviet trade during the 1970s and early 1980s was funded by windfall profits from resource exports. As Figure 9.1 illustrates, since 1985 the situation has become more complicated. During 1986 the price of oil fell from $ 30 a barrel to around $ 10. As a consequence the Soviet Union saw its oil revenues fall from $ 16.1 billion in 1984 to $ 7.07 in 1986. The situation with exports to CMEA was somewhat different with prices being based on a five-year moving average, but by 1989 the Soviet Union had a deficit in its trade with Eastern Europe. During 1985 production problems made it difficult to increase the volume of oil exports. However, according to Soviet statistics the total volume of oil exported (including to the CMEA) increased from 117 million tons in 1985 to 129 million in 1986 and 137 million in 1987. Between 1985 and 1987 the volume of oil exports to the OECD rose 22 per cent, while the revenue from those exports fell 38.8 per cent. In 1989 this strategy was further complicated by a 3 per cent fall in Soviet oil production, which resulted in an 18.8 per cent drop in oil and oil product exports to the OECD (Gorst, 1990). Fortunately for the Soviet Union, oil price increases during 1989 compensated for the decline in the volume of oil exports, resulting in a small increase in the value of hard-currency oil exports. In the first half of 1990 the situation showed little sign of improvement, but the Iraqi invasion of Kuwait and the subsequent increase in oil prices provided the Soviet Union with a much-needed short-term injection of foreign currency.[6] The most immediate problem is that a crisis in the Soviet oil industry, caused by mismanagment rather than a lack of resources, is likely to lead to further reductions in the volume of oil available for export. In the longer run the prospect of higher energy prices may persuade Western oil companies to increase investment in the Soviet oil industry. In sum, the energy price rises during the 1970s provided the Soviet Union with an opportunity to expand trade, particularly East-West trade, and to bolster a stagnating economy. But in the early 1990s there is no external panacea for the ills of the Soviet economy.

Foreign trade and Soviet economic performance

Centralized control made it possible for planners

to set export targets to finance imports to meet priorities set by national economic planning (Smith, 1987). It is only recently that the Soviet Union has recognized the need to expand imports regardless of the level of export earnings. Even now the traditional conservatism is reflected in the Soviet request for Western governments to provide credit to help resolve the payment problems that emerged during 1990.

The commodity structure of Soviet imports suggest three roles for foreign trade: first, imports of manufactured goods (particularly from Eastern Europe) have bolstered inadequate domestic production of consumer goods; second, imports of machinery and equipment have compensated for problems with the domestic innovation process and helped to sustain the development of priority sectors of the industrial economy; and third, imports of agricultural products have improved the diet of Soviet citizens and provided insurance against the inefficiencies of Soviet agriculture (it is not uncommon for 10–15 per cent of the domestic harvest to be wasted) and climatic uncertainty.

For the most part, the Soviet economy relies upon domestic inputs. The importance of foreign trade lies not so much in the volume of turnover, but that imports have served to remove bottlenecks and compensate for key weaknesses. This is particularly so in the case of imports of Western technology (Hanson, 1981; Cooper, 1985). Imports from the industrialized West represent a very small share of industrial investment within the Soviet economy; however, their impact has been amplified by the fact that they have been concentrated in particular sectors. The major recipients of Western machinery and equipment have been the chemicals and petrochemicals, oil and natural gas (particularly pipeline construction) and automotive industries. A number of case-studies by Western economists have examined the impact of Western technology transfer upon key sectors (see Holliday, 1984, for a review). The consensus among Western specialists would appear to be that in certain cases imports of Western technology have provided a 'quick fix' by enabling the introduction of new products as well as the expansion of existing production capacity; at the same time the systemic problems plaguing the Soviet economy have tended to reduce the impact of this 'imported innovation'. Since the mid-1970s the share of grain imports have increased, perhaps through recognition that it is easier to export oil and import grain than address the problems facing domestic agriculture. In 1989 the domestic grain harvest was 211.1 million tons and the Soviet Union imported 36 million tons (equivalent to 17 per cent of domestic production). The debate over bread prices in June 1990 revealed that every third loaf of bread baked in the Soviet Union was produced with imported grain. Given the symbolic importance of bread to the Soviet population, this is a severe indictment of Soviet agriculture, as well as a drain on foreign currency reserves.

However we measure it, foreign trade over the past fifteen years has been more important to Soviet economic performance than at any time during the Soviet period. In large part it is the central control of foreign trade that has enabled planners to use trade to address national economic problems. The decentralization of foreign trade presently under way will change this situation. To appreciate fully the implications of decentralization it is necessary to examine the regional distribution of Soviet foreign-trade participation.

East-West trade and Soviet regional development

There are no official data available on the regional distribution of Soviet trade with the industrialised West, so we have to use alternative sources of information to estimate the spatial distribution of export production and import utilization. In the case of export production, the commodity structure of Soviet exports can be compared with the pattern of regional industrial activity to arrive at estimates of the regional distribution of export production (McAuley, 1979; North, 1983). In the absence of detailed information on the consumption of imported consumer goods and agricultural products, we must assume that consumption follows the distribution of population. According to the preliminary results of the 1989 census, the European regions of the Soviet Union accommodate 66.1 per cent of the Soviet Union's population, Siberia and the Far East 11.2 per cent, Kazakhstan and Central Asia 17.2 per cent and the Transcaucasus 5.5 per cent. Geographical information on imports of Western technology can be obtained from the specialist Western business press dealing with East-West trade.[7] A database with information on over two thousand contracts, covering the period since the early 1970s, has been created to examine the regional distribution of Western technology transfer (Bradshaw and Shaw, 1989 and 1991).

A simple model of trade and regional development enables us to relate the commodity structure of East-West trade to the pattern of regional development in the Soviet Union (Figure 9.2). The starting point in this trade and development process is a demand for foreign currency that is met primarily by the sale of raw materials. The foreign currency generated by exports is then used to purchase imports of machinery and equipment, manufactured goods and agricultural products. Under the centralized system of management the transfer of funds between exporting and importing sectors was accomplished by the Ministry of Foreign Trade. As we shall see, the decentralization of the foreign trade system makes it all the more important to identify the regions that produce exports and the regions that consume imports.

The geography of foreign currency exports

Analysis of OECD trade statistics reveals that six commodity groups account for the majority of Soviet exports to the West: cork and wood (SITC 24), coal (SITC 32), oil and oil products (SITC 33), natural gas (SITC 34), diamonds (SITC 66) and non-ferrous metals (SITC 68). In 1984, before the fall in oil prices, these six commodity groups generated 87 per cent of exports to the OECD. Gold sales are not included in the OECD statistics but exports of gold have often been used to compensate for shortfalls in exports earnings. More recently the Soviet Union has deposited gold in Western banks to guarantee creditworthiness in the face of payment problems and a growing trade deficit. Exports of platinum-group metals are also an important source of foreign currency; these are reported in the non-ferrous metals commodity group.

Comparison of the commodity structure of exports and the geography of resource industries in the Soviet Union suggest that Siberia and the Far East are the most important regions for export production (Granberg and Rubenshteyn, 1989). The growth of oil and gas exports discussed earlier has been paralleled by a dramatic shift in the geography of energy production (see Chapter One). In 1975 Siberia and the Far East produced 30.7 per cent of Soviet oil and 39.8 per cent of Soviet gas. In 1989 these regions produced 67.3 per cent of the nation's oil and 68.1 per cent of its gas (Sagers, 1990a). Siberian oil and gas production is

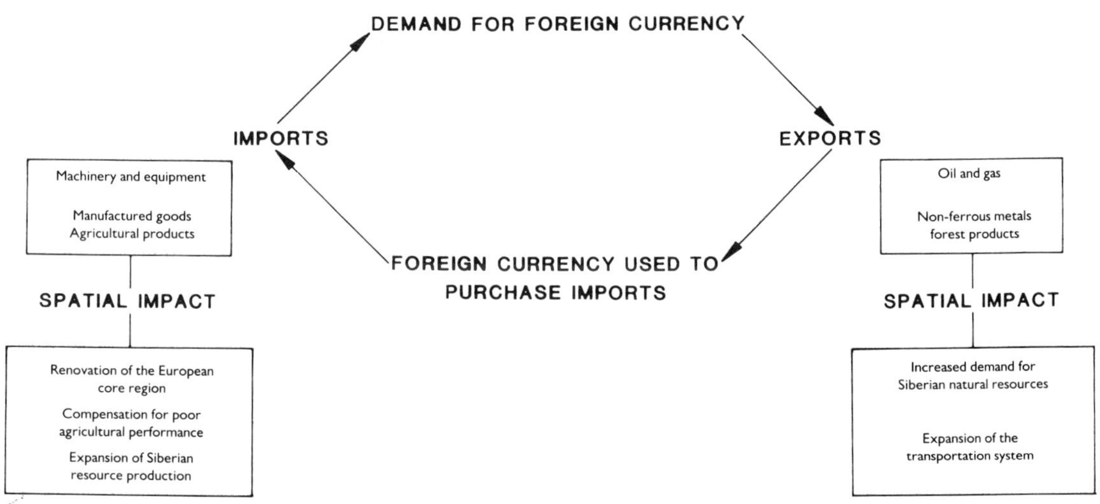

Figure 9.2 *The foreign trade and regional development process*

concentrated in Tyumen' oblast, within the West Siberian economic region, which in 1989 accounted for 99 per cent of the region's oil and gas production (see Chapter 6 for a discussion of Siberian development). Siberia and the Far East are also important locations for non-fuel mineral exports; the Noril'sk metallurgical combine is a major source of platinum-group metals; the hydro-powered smelters of the southern region of East Siberia produce 60 per cent of Soviet aluminium; the Far East is the only source of natural diamonds and the gold mines of the region are the traditional centre of Soviet production (in recent years Central Asia has also become an important centre for gold production).[8] A series of large-scale compensation agreements with Japan have also led to exports of Kuzbass and South Yakutian coal and forest products from the Far East (Bradshaw, 1990).

In the absence of information on regional foreign trade participation, one has to assume that a region's share of export production for a particular commodity is equal to its share of national production for that commodity. For example, if Siberia produces 67 per cent of national oil production, it is assumed to have contributed 67 per cent of Soviet oil exports. An analysis of the six commodity groups that dominate exports to the Soviet Union reveals that Siberia's share of exports has increased from 18.6 per cent in 1970 to about 53 per cent in 1985, the value of exports increasing from $ 474 million in 1970 to $ 11.2 billion in 1985 (Bradshaw, 1991). The vast majority of this growth is due to the development of the West Siberian oil and gas complex. In 1985 Siberian oil and gas exports were worth an estimated $ 9.7 billion and represented 46 per cent of Soviet exports to the OECD. Granberg and Rubenshteyn (1989, p.25) estimate that Siberian oil and gas exports generate one-third of hard-currency exports. Whatever the precise figure, it is likely that Siberia and the Far East are responsible for over half the Soviet Union's foreign currency export production.

Outside Siberia and the Far East, it is the other regions of the Russian Federation (RSFSR) that are likely to generate the bulk of the remaining export revenue. In 1989 the RSFSR produced 91 per cent of the nation's oil, 77.4 per cent of its gas and 55 per cent of its coal. If we repeat the analysis of energy exports to the OECD, but consider all RSFSR production, in 1985 energy exports from the RSFSR accounted for 67 per cent of all exports to the OECD. The European regions of the RSFSR also contribute to exports of forest products (see Chapter 7) and non-ferrous metals. In addition, the Lada car plant that produces export models is located at Tol'yatti in the Volga Economic Region within the RSFSR. In sum, the Russian Federation, in particular Siberia and the Far East, is responsible for the majority (probably over 75 per cent) of Soviet foreign-currency exports.

The geography of East-West technology transfer

The following discussion is based upon a database of Soviet-Western business contracts (Bradshaw and Shaw, 1991). Unfortunately it has not been possible to collect value data for all the contracts recorded, so the spatial distribution of technology transfer is described in terms of the total number of contracts located within a particular economic region. Information on the location of Western imports is available for about 75 per cent of the 2,300 contracts in the database (joint ventures are not included in this analysis). The sectoral and regional distributions of technology transfer are shown in Table 9.2 and the spatial distribution is shown in Map 9.1.

The oil and gas, chemical and automotive industries have been the most significant recipients of Western technology. However, there is considerable variation in the sectoral structure of technology transfer among the various economic regions. The data in Table 9.2 report the sectoral pattern for the Siberian (Siberia and Far East) and European regions of the Soviet Union. Within Siberia it is the resource industries that have been the major beneficiaries of technology transfer. The West Siberian oil and gas complex has received pipe and pipeline equipment to move energy to markets in the European Soviet Union and export markets in East and West Europe; however, little Western equipment has been used to maintain production in the oil fields. At the same time, the Northern Sea Route has been an important destination for imports of transport equipment, predominantly in the form of ice-breakers and ice-

Table 9.2 *Sectoral distribution of technology transfer by key region*

(per cent of total)

Sector	USSR	Centre	Volga	Siberia and the Far East
Oil and Gas	18.1	0.0	5.8	34.7
Other Energy	2.7	0.6	0.0	8.8
Forest	8.7	3.7	0.8	15.4
Chemical	17.8	21.3	25.8	12.1
Metals	7.8	1.8	2.1	6.1
Automotive	14.3	29.3	59.2	0.6
Transport	9.2	1.2	0.0	14.9
Food	5.7	7.9	1.3	3.6
Textiles	6.6	14.0	1.7	1.4
Light Industry	3.2	9.1	1.3	1.1
Other*	5.9	11.0	2.1	1.4
Total	100	100	100	100

Note:
*Other includes services and electronics.
Source: ESRC project database.

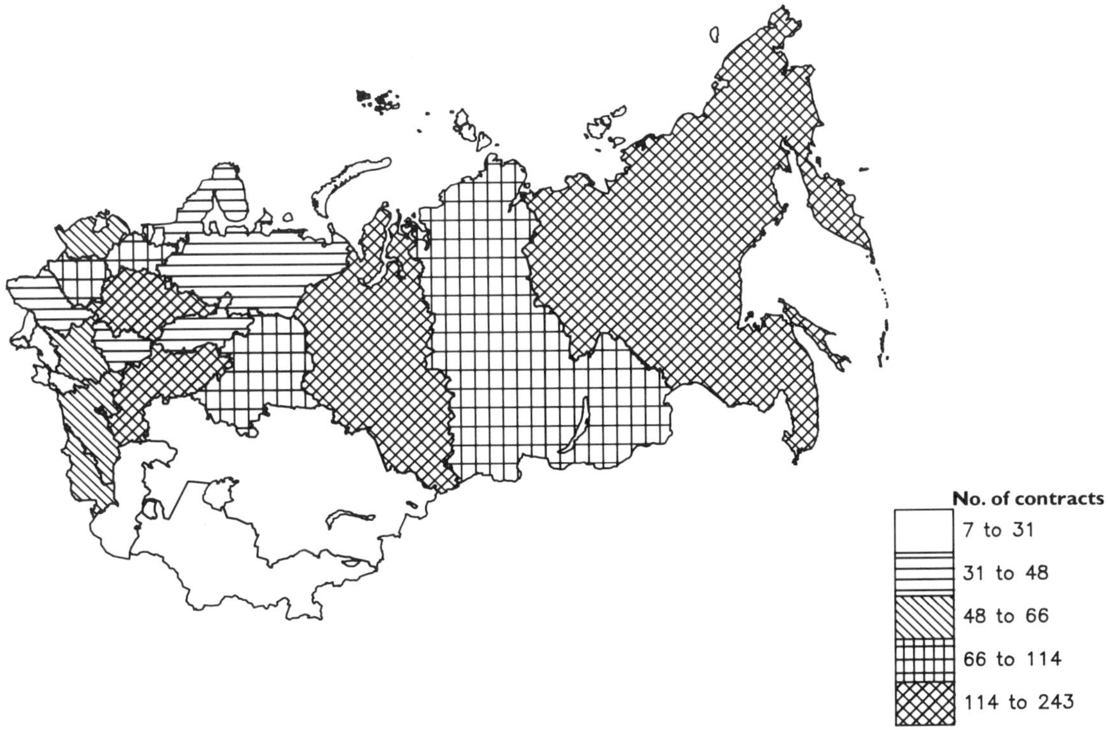

Map 9.1 *Regional distribution of Western technology (by economic region)*

strengthened shipping (the majority of this shipping has come from Finnish shipyards). The expansion of operations on the Northern Sea Route is also related to the development of oil and gas fields in the northern regions of West Siberia (Chapter 8). In the Far East, imports of machinery and equipment have aided in the exploitation of the region's forest resource. In the European regions the chemical and automotive industries have been the major recipients of Western technology, particularly in the Central and Volga economic regions. In the European Soviet Union the emphasis has been upon the expansion and improvement of the manufacturing industries; the resource industries are less important but the food, textiles and light industries have benefited from Western equipment. The macro-regional distribution of technology transfer shows that the European regions have received over three-quarters of Western contracts, but that there was an eastward shift in the late 1970s and early 1980s associated with the rapid development of Siberia's oil and gas resources. At the republican level the Russian Federation received over 65 per cent of all the contracts for which location data are available.

The geographical distribution shows a dual role for Western imports; in the European regions technology transfer has provided the 'quick fix' for key industries that is reported in the case-study literature; in Siberia and the Far East imports have provided specialist machinery and equipment to help the exploitation of the region's natural resources.

The trade and regional development process

Having examined the regional distribution of East-West trade in the Soviet Union it is now possible to make a number of conclusions about the relationship between foreign trade and regional development. First, the Russian Federation dominates East-West trade, as it does the domestic economy, accounting for about 75 per cent of foreign currency exports and receiving about 75 per cent of imports of Western technology.[9] Second, within the Russian Federation there is a transfer of foreign currency between the Siberian and European regions. It is the natural resources of Siberia and the Far East that generate the majority of the republic's foreign currency, but the European regions that benefit from the majority of imports. Thus, Siberia and the Far East (more specifically West Siberia) have generated a foreign-currency surplus that has been used to purchase imports for other regions within the Russian Federation, as well as the other republics in the Union. Third, the growth of Soviet foreign trade has aggravated the core-periphery relationship between the European and Siberian regions of the Soviet Union: imports have provided machinery and equipment to modernize the industrial base of the European regions, while exports have increased the demand for energy resources and raw materials, amplifying the resource-orientation of the Siberian economy. Finally, this regional transfer of foreign-trade revenues has taken place within the Ministry of Foreign Trade which has had All-Union responsibility. Thus the centralization of foreign trade has not only enabled the transfer of funds between sectors, but also between regions. This means that some sectors and regions do not benefit in proportion to the revenues they generate, while others receive imports they themselves cannot pay for. The reforms currently being introduced in the Soviet Union are changing both the sectoral and regional management of foreign-trade and the geography of foreign-trade activity is of increasing significance.

Opening the Soviet economy

The reform of the Soviet foreign-trade system began in August 1986 with the announcement of two decrees by the CPSU and the USSR Council of Ministers, one dealing with 'foreign trade measures' and the other with 'socialist cooperation measures' (Hanson, 1987). In January 1987 two further decrees were introduced on 'Comecon joint ventures' and 'Western joint ventures'. Together, these decrees, and subsequent amendments to them, form the legal basis for the foreign-trade reforms. The traditional system had granted the Ministry of Foreign Trade a monopoly over foreign economic relations. The reforms have dismantled this monopoly, reorganised the management of foreign trade and introduced new forms of international cooperation.

The intention of the reforms is to open the Soviet economy to the competitive pressures of the international market and to use foreign economic relations to promote the goals of perestroyka.

Foreign trade reforms

On 1 January 1987 the Ministry of Foreign Trade lost the monopoly over foreign trade it had enjoyed since the 1930s. For more than 50 years Soviet industrial ministries and enterprises had been required to operate through the Ministry of Foreign Trade and its organs. As of 1 January, twenty-two ministries and seventy-seven associations, enterprises and organizations were granted the right to carry out export-import operations directly on the foreign market. In late 1988 the right to direct-access to foreign markets was extended to most ministries, all the Union Republics and more than one hundred enterprises, inter-sectoral scientific-technical complexes and other organizations (Shmelev and Popov, 1990, p. 231). In December 1988 the USSR Council of Ministers decreed that as of 1 April 1989 the right to carry out export-import operations would be extended to all enterprises, associations, production cooperatives and other organizations with products that would be competitive on the international market. Thus in the space of two years the monopoly was replaced by a system allowing all internationally-competitive entities to participate in foreign-trade (probably about 1,800 organizations). The rapid devolution of the foreign-trade system has been paralleled by a reorganization of the management structure as well as a good deal of confusion (Geron, 1990).

The reorganization of the foreign-trade system has seen the amalgamation of the Ministry of Foreign Trade and the State Committee for Foreign Economic Relations to form the Ministry for Foreign Economic Relations (*Ministerstvo Vneshnikh Ekonomimicheskaya Svyazeye SSSR*). A number of the FTOs previously subordinate to the Ministry of Foreign Trade have been transferred to the appropriate industrial ministries, such as *Exportles* being transferred to *Minlesprom* (Chapter Seven), while other ministries created their own FTOs. Before the reforms the Ministry of Foreign Trade had forty seven FTOs; by the end of 1988 it had twenty five. Previously the FTOs of the Ministry of Foreign Trade controlled 95 per cent of imports from the West; by the end of 1988 its share had fallen to 40 per cent. This dramatic shift in responsibility has been brought about because the majority of the FTOs transferred to other ministries are in the manufacturing industries. The new Ministry for Foreign Economic Relations retains control over exports of energy and raw-material products – and therefore the vast majority of Soviet foreign currency exports.

The sectoral differentiation of foreign-trade decentralization clearly illustrates how reform measures have distinct spatial impacts. An unintended consequence (perhaps) of these reforms is that the manufacturing industries of the European Soviet Union benefit from the devolution of foreign-trade rights, while the resource exports of Siberia and the Far East remain under Moscow's control. In late 1990 a Union Republic Currency Committee was established to manage the distribution of export revenue. Adjustments have also been made to foreign-currency retention quotas which, by allowing an exporting enterprise to retain a proportion of the hard currency generated by its exports, aim to encourage export production. Traditionally, the retention quotas have been structured to promote the export of higher-value-added goods. Thus the quota has been zero for some raw materials and as high as 80 per cent for the production of sophisticated machinery (UNCTAD, 1988, p. 7). This discriminated against regions such as the Far East whose economic structure is biased towards raw material production. There are signs that this is changing, as 1991 retention quotas will allow resource industries to retain a small portion of their hard-currency export revenue (Business USSR, 1991). The issue of export revenues has been at the centre of a number of labour disputes in the energy industry. Planners and politicians in Moscow wish to use the potential rewards of foreign-trade participation to encourage the modernization of Soviet industry, but to date they have failed to reward the resource industries responsible for generating the majority of the export revenue.

In addition to the Ministry of Foreign Economic Relations, a 'superministry' entitled the 'State Commission for Foreign Economic Relations' has been created. The State Commission is a permanent body of the USSR Council of Ministers and its role is 'to formulate and implement the country's foreign economic strategy so as to enhance its potential contribution to *uskoreniye* (accelerated economic growth)' (Ivanov, 1987, p. 197). The membership of the commission includes representatives of all the major agencies that regulate or are involved in foreign-trade activities. Under this new management structure the central ministry and state commission are more involved in promoting and facilitating foreign trade, rather than acting as an intermediary to isolate Soviet enterprises from the international market. An individual enterprise is now faced with a number of different channels for gaining access to foreign markets: first, it may use the traditional route of engaging the services of an FTO subordinate to the Ministry for Foreign Economic Relations; second, the enterprise may use an FTO subordinate to another ministry (perhaps one related to the enterprise's activities) or an FTO created by another organization; finally, the enterprise could create its own FTO and conduct its own negotiations. Involvement with the central ministries has the benefit of experienced negotiators (although some of the new FTOs have had problems finding qualified staff) and the greater likelihood of official credits from The Foreign Trade Bank (*Vneshekonombank*). Going it alone is far more risky as it depends upon the marketing abilities of the enterprise (which are seldom adequate) and may not provide access to state foreign-currency reserves (which could result in payment problems).[10] The reorganization of the foreign-trade system and the devolution of foreign-trade rights have introduced uncertainty into a system that for decades was renowned for its stability. Soviet enterprises are now having to find their way in a new economic environment. One way of gaining access to foreign expertise and revenue is to create a joint venture.

Joint-venture legislation

Two decrees allowing Soviet enterprises to establish joint ventures with foreign companies were introduced on 13 January 1987, one dealing with joint ventures with COMECON partners and the other with Western partners. In late 1990 75 per cent of the joint ventures registered with the Ministry of Finance were with Western partners. Given the practical demise of COMECON and the conversion of Soviet-East European trade to hard currency, it is likely that this distinction will be of less significance in the future. For our purposes, the following examination of joint-venture legislation focuses on Soviet-Western joint ventures.

The decree (No. 49 of the USSR Council of Ministers) allowing the establishment of joint ventures with capitalist and developing countries identified the goal of joint ventures as being:

> to satisfy more fully domestic requirements for certain types of manufactured products, raw materials and foodstuffs, to attract advanced foreign equipment and technologies, management expertise, additional material and financial resources to the USSR national economy, to expand the national export sector and to reduce superfluous imports (ECE, 1988, p. 183)

For the Soviet Union the principal goals for joint ventures are: technology transfer, the improvement of the structure of exports (meaning a reduction of the reliance upon energy and raw materials) and import substitution. For prospective Western partners the principal goals are access to resource wealth (that can be easily sold on the world market) and to a potential market of 280 million people. The problem with the Soviet market today is that those 280 million Soviet citizens are paid in non-convertible roubles. For Western companies, the single greatest problem is how to repatriate profits from operations in the Soviet Union. As of 1 January 1991, joint ventures were able to convert roubles into hard currency by participating in foreign-currency auctions in Moscow; however, in late-1990 the auction exchange rate was considerably lower than the commercial rate. The evolution of joint-venture legislation reflects the inherent conflict in Soviet and Western goals and an attempt to reach a compromise that promotes domestic reform and at the same time provides profitable opportunities for Western investors. So far it has been a process of trial and error and many more adjustments will probably be introduced.

The initial legislation of January 1987 required that 51 per cent of the joint venture be owned by the Soviet partners and the chairman of the board and general director had to be Soviet citizens. The Soviet share of the foundation capital was to be paid in roubles on the basis of an agreed exchange rate reflecting world prices. The Western contribution would be calculated at the official exchange rate (which leaves joint ventures vulnerable to devaluation). The decree also included regulations governing property rights and liability. Joint ventures were exempt from state plans and permitted to engage in foreign trade through the FTO system. Hard-currency profits could be repatriated only when they exceeded the value of foreign-currency imports. This measure stops joint ventures being used to gain direct access to the domestic market. After its creation, the joint venture would be granted a two-year tax holiday, after which profits would be taxed at 30 per cent (after deductions to their reserve and re-investment). An additional 20 per cent tax would be levied on profits transferred abroad. Clearly these regulations are rather restrictive and the ensuing Western response was lukewarm. During 1987 only twenty three joint ventures were registered.

On 17 September 1987 a number of amendments were introduced to induce greater Western interest. Special allowances were introduced to reward import substitution; profits from sales to the state can now be repatriated, regardless of the level of imports, if they involve import substitution. The two-year tax holiday now dates from the moment of showing a declared profit, rather than the date of establishment. Further allowances increased the freedom allowed joint ventures: they no longer have to trade through FTOs and are allowed to buy and sell in the domestic economy in any currency. Despite these changes, in 1988 only 160 joint ventures were registered, suggesting that these revisions were not considered sufficient by many prospective Western partners.

On 2 December 1988 a more substantial set of revisions were introduced that further improved the status of joint ventures. Among the most significant concessions are: the Western share can now be greater then 50 per cent, the general director can be a non-Soviet citizen, the joint venture now has a greater control over its workforce, goods imported for investment purposes face minimal custom duties and local costs can now be paid in roubles. A differential taxation system was introduced to encourage the establishment of joint ventures in particular sectors and regions. The level of taxation on profits is reduced to 10 per cent (instead of 30 per cent) for joint ventures located in the Far East. Similarly, joint ventures producing consumer goods, medicines, medical equipment, high-technology goods and goods of significance to the national economy are now also subject to reduced taxation. The initial Western response to these concessions was positive, with over 1,000 joint ventures being registered with the Ministry of Finance during 1989. During early-1990 the rate of registration was 80-100 a month. However, by the end of 1990 the rate of joint-venture registration slowed, reflecting Western concern about the scale of Soviet economic problems and the introduction of alternative forms of investment, such as joint-stock companies and wholly-owned subsidiaries (Morgachev and Chaplin, 1990).

On 7 March 1990 a further set of amendments were proposed but their official status remains unclear. They included the setting of a 200,000 rouble minimum for total investment, with the Western share a minimum 100,000. A time-limit was introduced within which the Western partner must deposit its share of capital in the relevant Soviet banks. There is also the stipulation that joint ventures must add value to a product and are no longer allowed to engage in counter trade (one of the few ways of repatriating profit). Western business is bound to view these new measures as restrictive: they close a number of loopholes that joint ventures were already exploiting. In addition the operational environment is continually changing due to new All-Union laws and decrees on exchange rates, foreign investment and property rights (Louis et al., 1990). The continual adjustment of the conditions governing the operations of foreign companies only serves to complicate an already difficult situation. To assess the progress made so far, the following section examines the activities of Soviet-Western joint ventures.

Soviet-Western joint ventures

By the spring of 1991 an estimated 3,000 joint ventures had been registered with the Ministry of

Finance in Moscow. However, the fact that a joint venture has been registered simply means that its creation has been approved by the Ministry of Finance: it does not mean that it physically exists. In the future the situation will be further complicated by the fact that joint-venture registration has been transferred to republican and local authorities. It is difficult to discover the actual status of registered joint ventures; business estimates suggest that about 400 (or 20 per cent) are actually operational. In many instances the Western partner has yet to deposit its share of the foundation capital in the appropriate Soviet banks. The Western business press is now reporting the failure of joint ventures and the cancellation of agreements, further evidence that problems in the Soviet economy are undermining investor confidence. It is likely that in the future the few successful joint ventures will be converted into Western-owned subsidiaries.

The data in Table 9.3 compare joint venture activity to participation in East-West trade. The most important joint-venture partners are West Germany, the United States and Finland. A number of Western nations have been more active in joint-venture creation than in commodity trade, while Japan is notable for its relatively low level of joint venture creation. The hesitancy of the Japanese reflects persistent political problems over the Kurile Islands (which remain unresolved despite Gorbachev's visit to Japan) as well as the less-than-satisfactory experience of Japanese companies with compensation projects in the Soviet Far East.

The majority of joint ventures are small operations: 60 per cent have a foundation capital of less than one million roubles. The small scale of joint ventures is explained by the types of activities in which they are engaged (Table 9.4). Manufacturing activities account for almost 50 per cent of the joint ventures registered and 60 per cent of the foundation capital. The majority of the other activities are related to the service sector. Within the manufacturing sector the production of office equipment and computers is the most popular activity. The creation of a joint venture in the service sector is relatively low risk when compared to setting up manufacturing facilities: the capital investment is low, there is huge unsatisfied demand in the major cities and in many instances payments for services can be made in foreign currency. Given the sectoral orientation of joint ventures it is no surprise that their regional distribution follows that of the population. Moscow alone accommodates 50.5 per cent of all joint ventures (Table 9.5 and Map 9.2). It is worth noting that the Baltic republics (especially Estonia) have been particularly active in negotiating joint ventures. Clearly they see them as a way of increasing their economic independence from

Table 9.3 *Comparison of Western joint-venture activity and participation in Soviet foreign trade*

Western Partner	JVs[1]	Per cent	Trade 89[2]	Per cent	JV Trade
West Germany	263	18.1	6.6	17.8	1.0
United States	214	15.1	3.4	9.2	1.6
Finland	179	12.6	3.9	10.5	1.2
Austria	111	7.8	1.4	3.9	2.0
Great Britain	101	7.1	3.2	8.7	0.8
Italy	120	8.5	3.5	9.6	0.7
Switzerland	71	5.0	1.4	3.8	1.3
Sweden	59	4.2	1.0	2.7	1.6
France	65	4.6	3.9	10.5	0.4
Japan	33	2.3	3.5	9.4	0.3
Share of Western	1216	85.7	31.8	86.1	1.0
Total Western	1418	100.0	36.9	100.0	
Share of total		75.3		26.1	2.9
USSR total	1884	100.0	140.9	100.0	

Notes:
1. Joint ventures registered by 1 October 1990.
2. Based on total trade turnover (in billion roubles) from Soviet trade statistics.

Table 9.4 *Soviet joint-venture activities (as of 1 October 1990)*

Industry	Number	Per cent	Capital[1]	Per cent
Agriculture, hunting and forestry	24	1.3	53.4	1.2
Fishing	9	0.5	52.6	1.1
Mining and quarrying	4	0.2	3.1	0.06
Manufacturing	934	49.6	3039.9	66.2
Purification of water	3	0.2	7.6	0.2
Construction	86	4.6	136.4	3.0
Wholesale and retail trade	61	3.2	190.1	4.1
Hotels and restaurants	128	6.8	344.7	7.5
Transport, storage and communications	56	2.9	57.4	1.3
Financial intermediation	5	0.3	129.7	2.8
Real estate	4	0.2	5.7	0.1
Renting of machinery and equipment	23	1.2	19.2	0.4
Computer and related activities	109	5.8	87.5	2.0
Research and development	11	0.6	35.4	0.8
Other business activities	257	13.6	156.6	3.4
Education	19	1.0	10.4	0.2
Health and social work	45	2.4	116.0	2.5
Sewage and refuse disposal	10	0.5	12.9	0.3
Cultural and sporting activities	81	4.3	87.9	1.9
Other services	5	0.3	2.2	0.04
Other	10	0.5	42.1	0.9
Total	1884	100	4590.6	100

Note:
1. Million roubles.

Source: UN European Commission for Europe, database on joint ventures.

the Soviet Union.

There can be no doubt that joint ventures have seized the attention of the Western media. Indeed, establishing a joint venture may even be seen as an effective marketing tool in Western markets. However, McDonald's and Pizza Hut are not going to solve the problems facing the Soviet economy. Joint ventures represent a very small share of foreign trade in an economy that is largely self-sufficient. During 1989 the combined export and import activities of Soviet-Western joint ventures accounted for a mere 1.5 per cent of the total trade between the Soviet Union and the West (calculated from Geron, 1990). A more realistic appraisal of joint ventures must be delayed until more of them are actually operational. Although over 2000 have been registered, it is unlikely that more than a third of those will actually succeed. There have been a number of large joint ventures proposed but their experiences suggest that without substantial changes in the domestic economy, joint ventures will do little to open up the Soviet economy. The episode of the multi-billion dollar petrochemical projects in West Siberia is instructive.

In 1988 the Kremlin approved the development of five petrochemical projects in Tyumen' oblast in West Siberia at Surgut, Novyy Urengoy, Tobol'sk, Nizhnevartovsk and Uvat (Sagers, 1989). The total investment was later put at around $ 70 billion. Four of these projects were subsequently included in joint-venture proposals. Two US companies, Combustion Engineering and McDermott International, along with the Finnish state oil company Neste Oy were involved in proposals at Tobol'sk and Surgut. A project at Novyy Urengoy involved McDermott International and the Japanese companies Mitsui and Toyo Engineering. A consortium of Japanese companies was also involved in a project to

Table 9.5 *Regional distribution of joint ventures*

	Dec. 1988	Per cent	Oct. 1989	Per cent	Oct. 1990	Per cent
Armenia	3	1.6	8	0.9	14	0.7
Azerbaydzhan	2	1.0	6	0.6	11	0.6
Belorussia	3	1.6	16	1.7	38	2.0
Estonia	13	6.8	63	6.8	105	5.6
Georgia	7	3.7	29	3.1	54	2.9
Kazakhstan	1	0.5	8	0.9	14	0.7
Kirgiziya	0	0.0	0	0.0	1	0.1
Latvia	2	1.0	23	2.5	46	2.4
Lithuania	4	2.1	10	1.1	18	1.0
Moldova	2	1.0	8	0.9	21	1.1
RSFSR	140	73.3	689	74.2	1401	74.4
Moscow[1]	94	49.2	482	51.9	951	50.5
Leningrad	13	6.8	80	8.6	177	9.4
Other RSFSR	33	17.3	127	13.7	273	14.5
Tadzhikstan	0	0.0	2	0.2	3	0.2
Turkmenistan	0	0.0	1	0.1	1	0.1
Ukraine	9	4.7	57	6.1	133	7.1
Uzbekistan	2	1.0	7	0.8	20	1.1
Unknown	3	1.6	2	0.2	4	0.2
USSR total	191	100	929	100	1884	100

Note:
1. Figures for Moscow and Leningrad include joint ventures in the surrounding region.

Sources: 1988: ESRC project database.
1989 and 1990: UN European Commission for Europe, database on joint ventures.

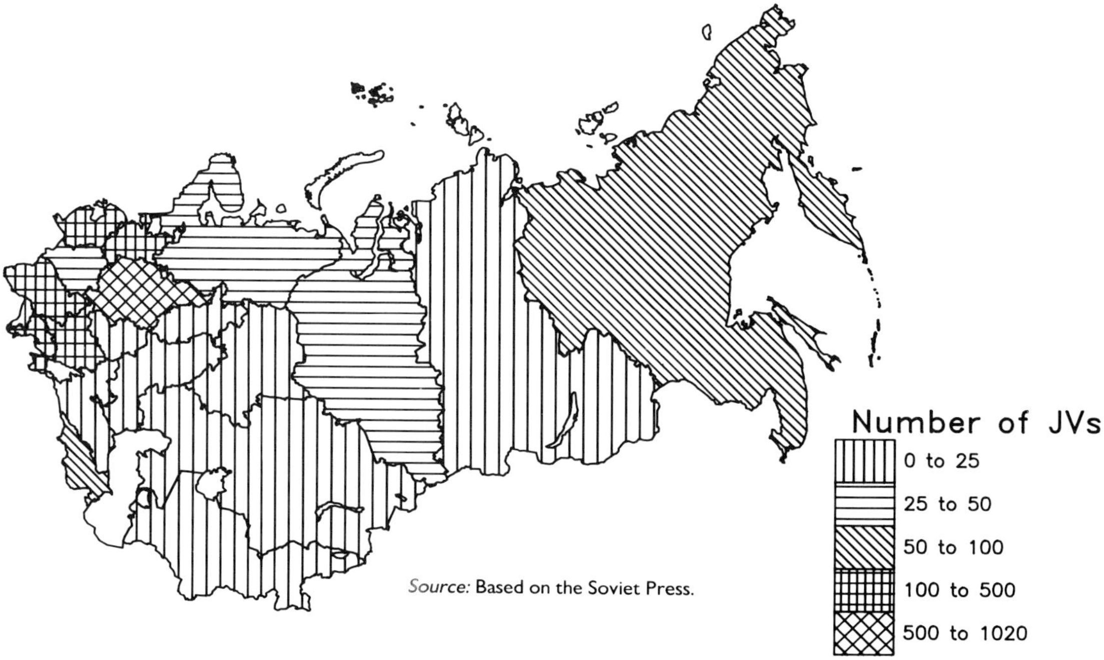

Map 9.2 *Regional distribution of joint ventures (by Economic region, as of 4 October 1990)*

construct a petrochemical complex near Nizhnevartovsk. The aim of these projects is to make greater use of local hydrocarbon resources, much of which are simply flared off, as well as to attract foreign investment and generate export revenue. Local opposition groups protested that the environmental consequences of the projects would turn Tyumen' into 'a world latrine' (Sagers, 1990b). Specialists in Moscow and Siberia questioned the wisdom of so many large projects in an area with such an under-developed economic infrastructure. Two Academy of Sciences' commissions supported the concept underlying the projects, but recommended that the initial phase be limited to Tobol'sk and Surgut. The Politburo approved projects at Tobol'sk and Surgut and in November 1989 urged that work be speeded up. Negotiations were further complicated when McDermott International pulled out of the project. Despite this setback, the joint venture 'Wespec' was registered in May 1990 with a foundation capital of $660 million to build and operate a petrochemical complex at Tobol'sk. Of the five mega-projects proposed for Tyumen' oblast, Wespec seems the most likely to go ahead and the first phase, worth $ 2 billion, is due to start soon. The financial status of the other projects remains unclear as the Soviet government refuses to give credit guarantees, insisting that joint ventures should be self-financing. The project at Surgut has been postponed. These petrochemical complexes were supposed to illustrate the benefits of international cooperation; instead they are highlighting the problems associated with large-scale Western investment in the Soviet economy. It is now clear that the internationalization of the Soviet economy cannot take place in isolation from and ahead of a radical economic reform. The future for the joint venture looks bleak. A combination of economic and legislative uncertainty and intervening opportunities is likely further to reduce joint venture activity. At the end of 1990 the Soviet business weekly *Commersant'* concluded:

> It now appears clear that foreign investment in the Soviet Union through joint ventures has not and probably will not translate into anything more than an asterisk on a chart of a declining Soviet economy (Morgachev and Chaplin, 1990, p. 8).

Systemic reform and foreign trade

It is clear that perestroyka is in serious trouble. Despite Gorbachev's foreign policy successes and his popularity in the West, five years of political turmoil and piecemeal reform measures have brought the Soviet Union to the brink of disintegration. The problems facing Soviet foreign trade – a growing balance-of-payments deficit, a backlog of unpaid bills, declining credit worthiness and a lack of investor confidence – are in large part a product of the failures of perestroyka. With the benefit of hindsight, we must question the wisdom of decentralizing the foreign-trade system ahead of radical systemic reform. Soviet enterprises have been given the right to interact with the global economy but their everyday operations are still controlled by state orders and the domestic price system. The few enterprises that have succeeded in securing a Western partner face uncertainty as both the legal framework governing foreign investment and the domestic economic environment are constantly changing. Foreign-trade negotiations have become entangled in arguments between the central Soviet government and republican governments. For example, a $2 billion credit arrangement with DeBeers to market Soviet diamonds prompted the Russian Federation to introduce a law on economic sovereignty that claimed all resources within the RSFSR as republican property.[11] In Moscow itself joint ventures have been caught in disagreements between the Moscow City Soviet and district councils. In an economy of shortage, where the rouble has little or no purchasing power, access to foreign currency is an important way of obtaining goods and services. The fact that joint ventures are exempt from certain taxes and planning processes also makes them an attractive haven for Soviet enterprises. Joint venture activities are bound to come under the close scrutiny of the KGB with its new powers to inspect enterprises for corruption.

Faced with a profusion of laws and decrees, at All-Union and Republican levels, governing privatization, de-monopolization and marketization, it is almost impossible to formulate a picture of the contemporary legal framework for foreign investment in the Soviet Union. In such a context it is easy to see the appeal of the 500-day Shatalin Plan (*Perekhod k rynku*, 1990). For all its weaknesses, the plan did at least represent a program for radical economic reform. However, even before the compromise of the Shatalin and Abalkin plans was presented to the Soviet parliament, a number of republics announced their intentions of following a different course. In early 1991 it is clear that an All-Union Treaty clearly demarcating the division of responsibilities and rights between the central and republican governments is crucial to the successful implementation of any reform plan, however radical.

Conclusions: decentralization and internationalization

The state monopoly on foreign economic activity that existed before 1986 may have hampered the growth of Soviet foreign trade, but it did allow the transfer of convertible currency between sectors and regions within the Soviet economy. As we have seen, foreign currency raised by the export of Siberia's natural resources was used to purchase imports for consumption in the European regions of the country. The deintegration of Soviet foreign trade now means that many sectors and regions may lose access to foreign currency imports, while others may start to benefit from export revenues. This may be potentially good news for the oil-rich inhabitants of Tyumen' oblast who are already trying to negotiate their own exports contracts, but it may not be in the interests of the Soviet economy as whole. It is sectors such as manufacturing, machine-building, food-processing and the old industrial regions of the European Soviet Union that require radical restructuring and access to Western capital; but without central management of export revenue these sectors will have limited access to foreign currency. The Shatalin Plan recognized this problem and suggested that the state monopoly on strategic exports such as oil, gas, gold and diamonds be retained in the short term. The drafts of the All-Union treaty recognize the need for central control of gold and diamond reserves. In the short term, at least, it will be necessary to continue to mortgage Siberia's resources to restructure the industrial base of the European regions.

The decentralization of the Soviet economy has already encouraged an increased regionalization of foreign-trade activity. Under the centrally planned system there were two FTOs responsible for developing border trade: *Lenintorg* in the Leningrad region and *Dalintorg* for the Soviet Far East. Since 1983 when Sino-Soviet border trade resumed, there has been a rapid expansion in activity; border trade is now an important source of raw materials and agricultural products for the Soviet populations bordering China in the Soviet Far East and Central Asia. The expansion of border trade has not been without its problems: the unplanned movement of goods has created chaos on the railroads. During 1989 border trade between the Chinese provinces of Heilongjang, Jilin and Xinjiang was valued at 950 million Swiss Francs ($ 395 million).[12] Some of the contracts Chinese and Soviet enterprises have brought Chinese workers to the Soviet Union to compensate for local labour shortages. As the centralized control of foreign trade is relaxed, it is only natural that border regions will seek to exploit their geographical location. Thus we can expect further increases in local trade suited to local needs, rather than the dictates of the central planners in Moscow. The recent ban on barter trade specifically excludes Sino-Soviet border trade, recognition of the importance of such trade to the economy of the Soviet Far East.

There has already been considerable discussion of the possibility of setting up special economic zones in the Soviet Union: Leningrad oblast, Novgorod, Vladivostok, Sakhalin and Nakhodka in the Far East, along with Novgorod, have been identified as possible locations. Work on regulations governing special economic zones has been underway for some time, and in the Russian Federation local authorities are busy setting up the legal framework for the creation of special zones at Leningrad, Nakhodka, Vladivostok and Sakhalin. A November 1990 presidential decree on foreign investment acknowledged the need for special economic zones to promote inward investment. However, it is also likely that each republic will develop its own set of special economic zones, which will then compete for foreign investment. Whatever the eventual outcome of the reform process, it is clear that responsibility for foreign-trade activities is rapidly devolving to the republican, regional and city authorities. Individual republics and cities will then be competing for foreign investment in much the same way as local governments do in the West. For prospective foreign investors it is no longer possible to conduct negotiations in Moscow with the appropriate FTOs and industrial ministries; instead it is now necessary to deal with republican and local authorities and to assess the viability of enterprises in a number of different locations. There are already indications that Western companies are approaching the Soviet market in a different manner. Many Western companies are now opening offices in republican capitals. In such a situation, knowledge of local conditions may be a key to success. At the same time, it is clear that Soviet enterprises and regional authorities are going to have to play a more active role in the international market if the Soviet economy is to benefit from the internationalization of its economy. In the short-term the transition to republican control will add to an already confused situation. In the long-run the decentralization of Soviet foreign trade will result in a substantial change in the structure and geography of foreign-trade activity.

Notes

1. The support of the Economic and Social Research Council (ESRC) is gratefully acknowledged. Part of the research reported here has been funded by ESRC awards number D00232336 and R000232184.
2. It is difficult to see where all the necessary hard currency will come from. The economically independent Central European states will have to use convertible currency to pay for energy imports from the Soviet Union. In the first instance the Soviet Union will probably use deliveries to pay off its current debts to Eastern Europe. It has been estimated that this will bolster Soviet foreign currency export earnings by $ 11 billion. At the same time there is no guarantee that the Soviet Union will use this revenue to import manufactures from Central Europe: they are more likely to expand imports from the industrialized West. The result will probably be increased balance-of-payments deficits in Central Europe.
3. The situation is no clearer as there are now three exchanges rates: the official rate (£ 1= 1 rouble or $ 1 = 0.6 roubles) which is the long-standing rate; the commercial rate (£ 1 = 3.0 roubles or $ 1 = 1.8 roubles); and the tourist rate (£ 1= 10 roubles or $ 1 = 5.6 roubles). In April 1991 the Tourist rate was further reduced to $ 1 = 27 roubles.
4. With the collapse of the CMEA this classification will soon be redundant.
5. The Soviet Union is now conducting bilateral negotiations with its ex-CMEA partners on the conditions for settling the outstanding debts. A combination of hard-currency payments and deliveries of raw materials will probably be used.
6. Third-quarter foreign-trade statistics show a surplus in Soviet trade with the West as a result of increases in oil prices due to the Gulf crisis. The third-quarter value of oil exports to the West was up 35.5 per cent compared to the third quarter of 1989.
7. The most useful sources include: *Business Eastern Europe*, *East European Markets*, *Soviet Business and Trade*, *East-West (Fortnightly Bulletin)* and *Moscow Narodny Bank Press Bulletin*.
8. For more information on Soviet mineral production and trade see Ballinger (1985), Hewitt (1986), Kaser (1984) and the section on the Soviet Union in the Minerals Yearbook.
9. Data in *Narodnoye Khozyaystvo v SSSR 1989 g* enable us to calculate the republican distribution of Soviet foreign trade with all partners, based on internal prices, for the year 1988. According to these data the RSFSR accounted for 69.2 per cent of imports and 70.6 per cent of exports in 1988. Nearly 87 per cent of total trade turnover is attributable to the three Slavic republics: the RSFSR, Ukraine and Belorussia.
10. Western business in now reporting that even the support of *Vneshekonombank* is insufficient to guarantee payment.
11. This may backfire as Tyumen' oblast is now saying that the region's oil resources belong to the oblast and not to the entire Russian Federation.
12. Trade between partners is conducted on a barter basis and the Swiss franc is used as an accounting unit.

Suggested reading

Geron, L. (1990), *Soviet Foreign Economic Policy under Perestroika*, Pinter, London.

Hanson, P. (1981), *Trade and Technology in Soviet Western Relations*, Macmillan, London.

Lavigne, M. (ed.) (1991), *The USSR and East Europe in the Global Economy*, Cambridge University Press, Cambridge.

Schaffer, M. E. (ed.) (1985), *Technology Transfer and East-West Relations*, St Martin's Press, New York.

References

Ballinger, P.R. (1985), 'Probability of continued Soviet mineral self-sufficiency', *Resources Policy*, 11, 3, pp. 160–76.

Bradshaw, M.J. (1991), 'Technology transfer and Siberian development', in *Modern Siberia: Social and Economic Development*, Wood, A. and Joyce, W. (eds), Routledge, London, forthcoming.

Bradshaw, M.J. (1990), 'Soviet Far Eastern trade', in *The Soviet Far East: Geographical Perspectives on Development*, Rodgers, A. (ed.), Routledge, London, pp. 239–68.

Bradshaw, M.J. and Shaw, D.J.B. (1989), *West-East Technology Transfer and Soviet Regional Development: An Introduction*, Working Paper No. 50, School of Geography, The University of Birmingham.

Bradshaw, M.J.and Shaw, D.J.B. (1991), 'East-West trade and Soviet regional development: continuity and change', in *The USSR and East Europe in the Global Economy*, Lavigne, M. (ed.), Cambridge University Press, Cambridge, forthcoming.

Business USSR, (1991), 'Soviet Retention Quotas: What are they for 1991?', *Business Eastern Europe*, 20, 9, pp. 65–6.

Cole, J.P. (1985), 'The geography of Soviet foreign trade, 1918–84: statistical tables and commentary', *Soviet Geography*, 26, 5, pp. 315–82.

Cooper, J. (1985), 'Western technology and Soviet economic power' in *Technology Transfer and East-West Relations*, Schaffer, M.E. (ed.), St Martin's Press, New York, pp. 80–110.

ECE (1988), *East-West Joint Ventures: Economic, Business and Legal Aspects*, United Nations, New York.

Gardner, H.S. (1982), *Soviet Foreign Trade: the Decision Process*, Kluwer-Nijhoff, The Hague.

Geron, L. (1990), *Soviet Foreign Economic Policy under Perestroika*, Pinter, London.

Gorst, I. (1990), 'Oil exports tumble', *Petroleum Economist*, August, pp. 23–4.

Granberg, A, Rubenshteyn, A. (1989), 'Uchastiye Sibiri vo vneshneekonomicheskoy deyatel'nosti: problemy i perspektivy', *Vneshnyaya torgovlya*, 8, pp. 25–9.

Gruzinov, V.P. (1975), *Upravleniye vneshney torgovli*, Mezhudnarodnoyye otnosheniya, Moscow.

Hanson, P. (1981), *Trade and Technology in Soviet Western Relations*, Macmillan, London.

Hanson, P. (1986), *Soviet Foreign Trade Policies in the 1980s*, Berichte des Bundesinstituts fur Ostwissenschaftliche und internationale studien, Koln, No. 41.

Hanson, P. (1987), 'Reforming the foreign trade system', *Radio Liberty Research*, RL 104/87.

Hewett, E.A. (1986), 'Minerals in East-West trade', *Resources Policy*, 12, 3, pp. 175–86.

Holliday, G. (1984), *East-West Technology Transfer II: Sectoral Case Studies*, OECD, Paris.

Ivanov, I.D. (1987), 'Restructuring the mechanism of foreign economic relations in the USSR', *Soviet Economy*, 3, 3, pp. 192–218.

Kaser, M. (1984), 'The Soviet impact on world trade in gold and platinum', in *The Soviet Impact of Commodity Markets*, Kostecki, M.M. (ed.), Macmillan, London, pp. 156–72.

Louis, N., Mamut, A., Charkham, R. and Warren, A. (1990), *Doing Business in the USSR*, Kogan Page, London.

McAuley, A. (1979), 'Impact and Importance of Regional Development for Soviet Trade: An Economic Assessment', in *Regional Development in the USSR*, Oriental Research Partners, Newtonville, Mass., pp. 277–86.

Morgachev, S. and Chaplin, V. (1990), 'JV Registration: The Year in Review', *Commersant'*, 31 December, pp. 8 and 10.

North, R.N. (1983) 'The impact of recent trends in Soviet foreign trade on regional development' in *Soviet Natural Resources in the World Economy*, Jensen, R.G., Shabad, T. and Wright, A.W. (eds), University of Chicago Press, Chicago, pp. 77–123.

Perekhod k rynku, (1990), Arkhangel'skoye, Moscow.

Sagers, M.J. (1989), 'Joint Ventures in the Soviet energy sector', *PlanEcon Report*, 5, 40, pp. 1–11.

Sagers, M.J. (1990a), 'Review of Soviet energy industries in 1989', *Soviet Geography*, 31, 4, pp. 280–314.

Sagers, M.J. (1990b), 'Plans for West Siberian petrochemical complexes finalized', *Soviet Geography*, 31, 3, pp. 224–7.

Shmelev, N. and Popov, V. (1990) *The Turning Point: Revitalizing the Soviet Economy*, I.B. Tauris & Co. Ltd., London.

Smith, A.H. (1987), 'Gorbachev and the World – the economic side' in *The Soviet Union under Gorbachev: Prospects for Reform*, Dyker, D.A. (ed.), Croom Helm, London, pp. 126–63.

UNCTAD (1988), *USSR: New Management Mechanisms in Foreign Trade*, United Nations, New York.

10

Conclusion: the geographical dimensions of reform

Michael J. Bradshaw

Introduction

The idea for this book was first conceived in 1989 when perestroyka was in full swing. The papers were first presented in a special session at the Meetings of the Association of American Geographers in Toronto in April 1990. At that time the Soviet leadership in Moscow was setting its sights on transition to a market economy. The final versions of the essays presented in this book were completed in early 1991. Between the optimism of the summer of 1990 and the spring of 1991, much has happened to the world and to the Soviet Union. The Gulf War has been fought and the United States is seeking to shape a 'New World Order'. Unfortunately for the Soviet Union, perestroyka has run aground, the Shatalin Plan has come and gone, reform-oriented members of the leadership have resigned or been replaced, independence movements in the Baltic Republics have been repressed, and conservative (anti-reform) forces seem to be gaining the upper hand. At the same time people are defying central government bans and protesting on the streets of Moscow, calling for Gorbachev's resignation. The timing of this apparent conservative reformation is unfavourable for the Soviet leadership as well as for policy-makers in the West, for there can be no doubt that the future role of the Soviet Union in any 'New World Order' is dependent upon successful systemic reform. If the Soviet leadership is unable to deal with inter-republican tensions and fails to implement a radical reform plan, then the Soviet Union will be unable to maintain its status as a superpower.

Five years of rhetoric have done little to change the reality of the Soviet economy. If anything, as many of the studies in this book illustrate, piecemeal and often ill-conceived reform has made a bad situation even worse. It may be that what is economically necessary is politically unacceptable. But the time has surely come when the political issues must be resolved to enable radical economic reform to happen. The Shatalin Plan faltered, in part, because it was predicated on a devolution of power to the Republics, a move the Soviet leadership still seems unable to make. Against this backdrop of rapid change and political turmoil it is all too easy to adopt a wait-and-see attitude. Geographers may be tempted to postpone their analyses of the spatial dimensions of reform until the economic and political conditions have stabilized. This book is based on the view that the geography of reform is not merely a future *outcome*, but part of the process of transition itself. A fundamental context to the direction reform is taking is the geographical legacy of the Stalinist extensive development strategy. It is only with an understanding of the processes that have shaped the Soviet landscape in the past that we can begin to consider the geographical dimensions of reform today and the possibilities of a new regional geography.

The Stalinist legacy

For most of the period between 1926 and 1985 the Soviet economy was controlled centrally in Moscow; the only significant deviation was Khrushchev's 1957 *Sovnarkhoz* administrative (regional economic council) reforms. This experiment was short-lived and central control firmly re-established in 1965. Since then the central planning system has favoured national interests over local needs. The dominance of the Communist Party at all levels of decision-making meant that Party goals governed the development of Soviet society. Bureaucrats in this system gained access to power and privilege through loyalty to the Party and not the government or republic. The resultant system of Party and state served to centralize power with the Politburo and Party officials in Moscow. Within this system the responsibility for economic planning and plan implementation rested with GOSPLAN, the State Planning Agency, and the All-Union Industrial Ministries. All the major industrial sectors of the economy are controlled by All-Union Ministries in Moscow; their jurisdictions transcend those of the Republican Ministries who are responsible for the more service-oriented industries. As a result there exists a 'dual' economy within each Republic, with heavy industrial sectors under the direct control of Moscow and service activities locally controlled. (To confuse the situation further, there are also some sectors of the economy under dual subordination.) The balance between central and republican control is strongly influenced by the economic structure of each republic. A high proportion of heavy industry is likely to result in greater central control, while regions specializing in light industry or agriculture will have more local control. This relationship between industrial structure and the distribution of control is a key factor in determining the regional consequences of investment strategy. The dominance of the industrial ministries has been further enhanced by the planning system which sets targets and allocates resources primarily by industrial sector rather than by economic region or republic. Within this planning environment individual ministries concentrate their energies on achieving their own sectoral goals (which include plan fulfilment) and emphasize linkages within their own branch sector, rather than linkages outwards to other sectors. This tendency to internalize linkages leads to 'departmentalism' whereby ministries create their own empires to provide services such as housing, transportation and construction. As Denis Shaw's study of the Soviet city illustrates, this makes local planning difficult as the local-planning authorities have no control over the All-Union Ministries; it also leads to wasteful duplication of facilities and irrational transportation shipments as ministries use their own internal suppliers rather than local sources. As a result, at the local level major industrial enterprises may be poorly integrated into the local economy. Such distortions are made all the more possible by the dominance of physical planning indicators, such as plan fulfilment, over price and profit. The inadequacies of the pricing system make it impossible to use financial measures to penalize waste and reward efficiency. More often than not, security of supply is more important than cost considerations.

The cumulative consequences of this centralized planning system are that republican and local authorities have had limited control over economic activity within their region. Obviously, local authorities and their representatives can lobby the central policy-makers, but the interests of the Industrial Ministries, particularly those associated with the Military-Industrial Complex, tend to dominate. The regional economies of the Soviet Union, particularly within the RSFSR, are the result of a special form of external control.

The dominance of sectoral planning over spatial regional planning means that insufficient thought has been given to the consequent spatial distribution of economic activity. There exists no viable industrial location strategy (Pallot and Shaw, 1981). The primacy of sectoral planning and the desire to concentrate industrial activity in very large plants, supposedly to gain economies of scale, results in a high level of monopoly power within Soviet industry. Analysis in the Shatalin Plan revealed that within the machine-building industry almost 80 per cent of production is monopoly production. In addition, according to Goskomstat data, in 209 of the 344 most significant commodity groups in industrial production, the output of the largest production enterprise accounts for more than 50 per cent of the total volume of production for that product; for 109 enterprises that output exceeds 90 per cent (*Perekhod k rynku*, 1990, pp. 77–8). Such a degree of concentration would not be permitted in the West and it is no surprise that many enterprise managers in the Soviet Union show little interest in product innovation.

Unfortunately, location data are not available to chart the regional distribution of monopolies. But the implications are clear: if one plant produces 90 per cent of a particular product, the region within which that plant is located also has a monopoly within the Soviet market. Thus an unforeseen consequence of Soviet industrial policy is high levels of regional specialization and interdependence (McAuley, 1991). At present much of this interregional trade is internalized within the linkages of industrial ministries. However, if control were to be devolved to republican authorities and the industrial ministries were disbanded, a potentially difficult situation would arise. The erosion of central control has already led to widespread barter trade between enterprises as they seek to reorient their supplier networks. At the same time some regions would have monopoly control over key industrial activities: for example, when disturbances in Azerbaydzhan stopped the export of oil-industry equipment, oil production in Siberia suffered.

As we would expect, some republics are more trade dependent than others. The data in Table 10.1 show the share of exports and imports in republican production and consumption. The larger republics (RSFSR, Ukraine and Kazakhstan) are relatively self-sufficient, but overall the

Table 10.1 *Share of all imports and exports in the total volume of production and consumption by Soviet republic in 1988*

Republic	Share of outgoing products (including exports) in total production (%).	Share of incoming products (including imports) in total consumption (%).
RSFSR	12	15
Ukraine	16	17
Belorussia	27	25
Uzbekistan	20	22
Kazakhstan	12	19
Georgia	26	28
Azerbaydzhan	26	22
Lithuania	22	27
Moldavia	25	28
Latvia	24	27
Kirgizia	19	26
Tadzhikistan	20	27
Armenia	26	31
Turkmenistan	22	24
Estonia	24	29

Source: Narodnoye Khozyaystvo SSSR v 1989 g (1990), Finansy i Statistika, Moscow, p. 636.

level of interregional dependence appears to be high (Belkindas and Sagers, 1990, p. 631). These figures hide considerable sectoral variations in interregional dependence. Not surprisingly, the level of regional specialization is particularly high in the resource-oriented and heavy industrial sectors of the economy, while there is a greater degree of regional self-sufficiency in food and consumer-goods production which are the responsibility of the union republic ministries. Soviet energy production and heavy industry are highly concentrated within the Slavic republics of the Soviet Union. Obviously the fact that the RSFSR accounts for 76.2 per cent of Soviet territory is a key factor and within the RSFSR itself there is considerable regional specialization. The eastern regions of the Republic are the major sources of natural resources, while the Republic west of the Urals contains the heart of the Soviet manufacturing industry. According to unpublished Goskomstat data on industrial employment, in 1985 the European regions of the RSFSR accounted for 49.5 per cent of total industrial employment. The same data-set indicates that the European regions of the Soviet Union (including the Baltic republics) accounted for 74.5 per cent of industrial employment. Analysis of inter-republican economic relations suggests that the RSFSR, the Ukraine, Belorussia and Azerbaydzhan have been long-term net donors to the All-Union economy, while Uzbekistan, Kazakhstan and Tadzhikistan have been long-term recipients (Belkindas and Sagers, 1990, pp. 638–9). As the analysis in Chapter Two illustrates, economic development trends reveal a similar pattern: higher per capita values and rates of growth of national income produced have been typical of northern and western parts of the Soviet Union, whereas the south and east, especially the Central Asian republics, have exhibited indices well below the All-Union mean (see also Bond et al., 1990, p. 725). These analyses suggest that the industrial structure of the individual republics and economic regions, something over which individual republics have had limited control, has been a key factor in determining their status within the All-Union economic system.

In sum, a high degree of regional specialization and interdependence are the unforeseen outcome of sectoral planning, which to date has been managed within the central planning system. This regional geography is the product of the dominance of the centre over the regions. However, the balance of power is shifting as the republics gain greater control over their destiny. In the long term this devolution of power will bring with it a new regional geography: at present it is a major source of tension and conflict within the Soviet system.

The tensions of transition

If, as I have argued above, the centrally planned system produced a particular spatial distribution of economic activity, then systemic transformation must involve the construction of a new regional geography. While it is relatively easy to prescribe what a market-oriented Soviet economy should be like, the key issue at present is how to manage the process of transition (*perekhod*). Any analysis of the tensions of transition must recognize the interdependence of political change and economic reform, and the constraints provided by the existing geographical system. Figure 10.1 provides a framework for considering some of the key issues raised by the process of transition.

At the centre of the present crisis in the Soviet Union is a clash of interests. President Gorbachev and the All-Union government presently headed by Valentin Pavlov profess to represent the national interests of the Soviet Union. The presidents of the constituent Union Republics, such as Boris Yeltsin of the RSFSR, represent their own local interests. The most visible expression of the conflict between the two levels of government has been the so-called 'battle of laws' whereby Republican Parliaments have passed declarations superseding All-Union laws. At the same time there is considerable variety in the goals of the Union Republics: the most radical, such as the Baltic Republics and Georgia, seek independence from the Soviet Union, while others, such as Kazakhstan, seek a rejuvenation of the Federal Union. These variations in attitude were clearly reflected by responses to the March 1991 referendum on the future of the Union: the Baltic Republics, Moldavia and Georgia refused to participate, while the Central Asian Republics all returned a 'yes' vote of over 90 per cent. Clearly, the progress of economic reform is dependent upon the resolution of the centre-republic conflicts and the building of a consensus between the All-Union and Republican governments; but it may well be that this consensus will have to be forged once the dissenting republics have left the Union.

CENTRALLY-PLANNED	MARKET-ORIENTED
NATIONAL	LOCAL
PARTY	STATE
CENTRALIZED	DECENTRALIZED
SECTORAL	SPATIAL
VERTICAL	HORIZONTAL
PHYSICAL	FINANCIAL

Figure 10.1 *The tensions of systematic transformation*

This confrontation between the centre and the republics has been promulgated by the erosion of the position of the Communist Party in favour of the state system, and is as much a struggle between Party and non-Party interests as the centre versus the regions. Previously the Communist Party served to maintain a consensus between the centre and the representatives of the republics, who were party members. Now many Republican leaders have denounced the Communist Party and formed their own political platforms. As Robert Kaiser shows in Chapter Three, the emergence of the state and the federal structure as the framework for reform has created further tensions, for the simple reason that the federal system is based upon nationalities. Thus nationalism has emerged as a cohesive force at the local level and has been given legitimacy through the federal state system. Furthermore, the emergence of strong republican governments with their own reform trajectories has served to undermine the process of economic reform.

Central to the process of reform is the desire to decentralize economic decision-making to the enterprise level, thereby reducing the role of the industrial ministries in the everyday running of the economy. In many instances industrial ministries have been consolidated to create 'superministries' charged with strategic planning; at the same time enterprises have been given greater control over their own operations. However, it is now clear that for this decentralization to be effective the operational environment must also change. At present enterprises remain subordinate to the ministerial system through the use of state orders (*goszakazy*); prices are still centrally controlled and enterprises are unable to raise the necessary capital and foreign currency to conduct their own modernization. Thus the enterprises are caught in a state of limbo, responsible for the outcome of their operations but unable to improve their production processes. Greater independence at the enterprise level would reduce the importance of vertical linkages and thus the tendency to departmentalize; at the same time it would generate horizontal linkages between suppliers in different sectors but in the same location. This would bring about a spatial reorientation of supplier networks within Soviet industry. The emergence of local and inter-republican barter trade is an indication that such an adjustment is already taking place. Even joint ventures, such as the UK-Soviet Femtek venture in the Ukraine which produces tampons, are having to pay for their inputs, in this case cotton, by providing products to their suppliers. Such barter trade is also an indication that the financial system is facing collapse. Failure to control the money supply and introduce effective price reform has undermined confidence in the Rouble. Enterprises would rather barter goods and services than pay for them with roubles. In such a situation, notions such as self-financing (*Khozraschet*) have little relevance when prices and money bear no resemblance to scarcity and purchasing power.

It is clear that for economic decentralization to work it must run in parallel with an increasing regionalization of the Soviet economy. This means that the republics must be given the necessary political powers and financial resources to effect change at the local level. But at the same time it is also apparent that there are certain sectors of the economy where central control must be retained or even strengthened. As Gorbachev's most recent draft of the Union Treaty suggests, the All-Union Government must remain responsible for certain aspects of finance and the strategic management of key sectors such as energy, transportation and defence. At the same time the central authorities may need greater powers to protect the environment and compensate for the social and economic problems created by the transition to a market-oriented economy. It is at this point that what is economically desirable runs against what is politically possible within the current environment. At present any decision to give the republics greater powers is resisted by conservative forces in the central government; any policy that seeks to maintain central control is rejected by the republican governments.

In any federal system it is natural for there to be tensions between the central federal government and the constituent governments: the recent history of Canada is a perfect example. In the Soviet Union the transition from a centralized federal system to a truly federal structure is presenting a particularly difficult set of problems. At issue is not only the distribution of responsibility but also the distribution of revenues and expenditures. It is here that the real tensions of transition and the legacy of the past emerge. The centrally planned system served as a vehicle to transfer resources between sectors and regions within the Soviet Union. Transition to a more decentralized market system will mean that revenues will have to be distributed between the central and republican governments; at the same time the richer republics will have to support the poorer republics. In most federal systems the distribution of income and the expenditure of revenue are key points of disagreement. In the Soviet Union the wealthier republics are the Slavic republics, while the poor republics are in Muslim Central Asia. Again, nationalism further complicates a difficult situation.

There are no easy answers to the problems that face Soviet society as it attempts the transformation from a planned to a market economy. But there are signs already that the process of systemic transformation and the tensions between the centre and the republics are producing a new regional geography.

The Soviet Union: a new regional geography

There are two senses in which this book has sought to examine the question of a new regional geography: first by considering the ways in which the contemporary regional geography constrains the progress of reform and the extent to which the spatial dimensions of reform are recasting the character and role of the republics and regions of the Soviet Union; and second, by developing new ways of thinking about regional change in the Soviet Union. In concluding, I wish to return to these two dimensions of Soviet regional geography.

Clearly, the story of perestroyka is far from complete. However, many of the geographical issues discussed in this book will be of lasting relevance because they form a crucial part of the Soviet reality that undermines the rhetoric of reform. The summer of 1990 may well be seen as a turning point in the progress of reform in the Soviet Union, not just because the Shatalin Plan was rejected but because it marked the start of a fundamental change in the nature of power relations within the political system. Since then the republican governments have begun to assert their own authority over the reform agenda. It is this re-orientation of power relations that has the potential to generate a new regional geography. The studies presented in this book highlight the confusion that has been created by centrally-decreed decentralization. It is now clear that progress towards a market economy is dependent upon coherent devolution of economic and political decision-making. However, the process of devolution will have to combat considerable inertia at all levels of Soviet society. Few doubt the need for change: the question is how to achieve a peaceful transition to a market-oriented economy. The issue of the day is no longer 'perestroyka', but 'perekhod' (transition).

The changes taking place in the Soviet Union demand new approaches to the study of geographical change. In the past, particularly during the Brezhnev period, the political system was taken for granted and the task of the geographer was to identify the geographical implications of central decisions. But even then, a rigorous regional geography required an analysis of the spatial consequences of the central-planning system (see, for example, Pallot and Shaw, 1981). In other words, to make sense of the regional geography of the Soviet Union it is necessary to appreciate the economic, social and political systems that produce the landscape. Now, perhaps more than ever, a political economy approach is essential to understand regional change in the Soviet Union. This collection of essays has considered the spatial consequences of central planning, assessed the reforms so far introduced, and the constraints imposed by the Stalinist legacy. This book has laid the foundations for a new regional geography. The challenge now is to come to terms with the process of transition.

References

Belkindas, M.V. and Sagers, M.J. (1990), 'Preliminary Analysis of Economic Relations Among Union Republics of the USSR: 1970–1988', *Soviet Geography*, 31, 9, pp. 629–56.

Bond, A., Belkindas, M.V. and Treyvish, A.I. (1990), 'Economic Development Trends in the USSR, 1970-1988: Part 1 (Production and Productivity)', *Soviet Geography*, 31, 10, pp. 705–31.

McAuley, A. (1991), 'Costs and Benefits of De-Integration in the USSR', *Most*, 2, forthcoming.

Narodnoye Kozyaystvo v SSSR 1989g. 1990, Finansy i Statistika, Moscow.

Pallot, J. and Shaw, D.J.B. (1981), *Planning in the Soviet Union*, Croom Helm, London.

Perekhod K Rynku, (1990), Archangelskoye, Moscow.

Index

Indexers Note: The references in *italics* are to maps and tabular material